biology in the contemporary world

MINICOURSE DEVELOPMENT PROJECT

Biological Sciences Curriculum Study □ Purdue University

S. N. Postlethwait, Director
Norris M. Ross, Associate Director
Ronald J. Wright, Associate Director

Ann Egan, Developer
Robert Egan, Developer
Ron Ladd, Developer and Consultant
Lane Lester, Developer
Keith Morrill, Developer
Stanley Nelson, Developer
August Wirkus, Developer
Bernadette Menhusen, Consultant
Kathleen Campbell, Administrative Assistant
Fay Ebrite, Administrative Assistant
Howard Manresa, Research Assistant
Beverly Barday, Secretary
Beverly Weiser, Secretary
James Thiele, Photographer
Evart Cable, Photographer
Jeanne Gorrell, Graphic Arts Technician
Judith L. Cardillo, BSCS Art Director
Louis Bruno, Project Art Director
Robert Wilson, Artist
Roy Udo, Artist
Arne Ludvigsen, Artist
John Evans, Artist
John Finn, Artist

Prepared by the BSCS with support from the National Science Foundation. Any opinions, findings, conclusions, or recommendations expressed herein are those of the authors and do not necessarily reflect the views of the National Science Foundation.

1976
W. B. SAUNDERS COMPANY
Philadelphia • London • Toronto

GRATEFUL ACKNOWLEDGMENT FOR ASSISTANCE IN THIS CLUSTER

Dr. J. Alfred Chiscon, Reviewer
Department of Biological Sciences
Purdue University
West Lafayette, Indiana

Dr. Peter Torino, Reviewer
Department of HPER
South Dakota State University
Brookings, South Dakota

Dr. Melvin Weinswig, Reviewer
Department of Pharmacology
University of Wisconsin
Madison, Wisconsin

Dr. Neal Groman, Reviewer
Department of Microbiology
University of Washington
Seattle, Washington

Dr. Robert Boyer, M.D.
Boulder Community Hospital
Boulder, Colorado

Dr. Merwin Moskowitz
Department of Biological Sciences
Purdue University
West Lafayette, Indiana

Ms. Michelle Baker
Colorado Mountain College
Glenwood Springs, Colorado

Biology in the Contemporary World ISBN 0-7216-6375-3

Printed in the United States of America. Press of W. B. Saunders Company. Library of Congress catalog card number 75-16822.

Last digit is the print number: 9 8 7 6 5 4 3 2 1

A LETTER TO STUDENTS

Albert Einstein, one of the greatest scientists and thinkers the world has ever known, once said, "The search for truth is more valuable than its possession." Certainly, I think that uncovering information for yourself through a variety of experiences is more interesting and more fun than simply sitting back and letting an instructor tell you all the answers before you've thought of the questions.

In these minicourses, you'll find a broad spectrum of experiences to help you pursue Einstein's "search for truth" in a subject which I think is the most exciting area of all—the study of living things. Those of us who have developed the minicourses believe that you really can learn biology and enjoy learning it—if you conscientiously involve yourself with the inquiries we have devised.

You have a great deal of control over exactly what you will see, hear, or do. On the one hand, you can proceed to a great extent at your own pace; you can turn the tape off if you need a break, repeat a segment of the tape if you miss a point, rework an exercise if something goes wrong, and take as much time as you need to work through the questions you'll be asked. On the other hand, you can create an unsatisfactory course for yourself; you can listen to the tape without involving yourself, you can "fudge" results of exercises, or you can decide not to ask for help when you're confused.

I hope you'll choose to involve yourself, because the effort will pay off. Very simply, you will learn biology. I know you will. I have used the audio-tutorial system in teaching for many years, and I have seen first-hand that nearly all students succeed as long as they let themselves participate fully.

Some students seem to hope that they can go through a course and pass it, without being changed by it. I hope you will not fall into that trap. Instead, please give yourself an opportunity to enjoy one of the greatest rewards of mankind—bringing about a change *in yourself.*

Paradoxical as it may seem, there is nothing so constant as change.

—Nehemiah Grew

Those of us who have been involved in developing the minicourses—and there were many of us—have tried to make sure that your efforts in learning will be directed toward the best and most current statements of biological ideas. To that end, we first consulted specialists from all over the world to determine just what the content ought to be. Then we translated these ideas into specific learning objectives. Your instructor has a list of these objectives for each minicourse, which he or she can modify to fit your specific needs and conditions. Quite possibly your instructor will choose to supplement our objectives with others to add a distinctive flavor to your course. We also developed more minicourses than you are likely to study; we hope that you can pursue your interest in some specific facet of biology with the aid of minicourses that your instructor chooses not to use in the mainstream of your course.

You do not educate a man by telling him what he knew not, but by making him what he was not.
—John Ruskin

One other aspect bears mentioning. We have organized each minicourse so that you will probably learn a good bit about how scientists tackle a problem. Each major section of a minicourse poses a question, and then you will proceed to answer the question much as a scientist would—building up some background information, comparing alternative hypotheses, and finally arriving at a likely answer.

I have always found biology to be an exciting inquiry into the nature of life. With your active participation, I think I can convince you to be just as excited as I am.

Sincerely,

S. N. POSTLETHWAIT

Director, Minicourse Development Program

BIOLOGY IN THE CONTEMPORARY WORLD: AN OVERVIEW

I. CANCER

Nearly one of every four persons living today will be treated for cancer at some time in life. Cancer confers no immunity by age, sex, social status, or any other arbitrary criterion. What is the current status of our knowledge of this kind of disease, and what hopes do we have for the future?

II. DISEASE AND TECHNOLOGY

A modern case history is used to illustrate the nature, causes, and control of diseases. The most common diseases are considered along with the methods available for their control and prevention.

III. THE BIOLOGICAL BASIS OF RACE

What ideas do you associate with race? Biologically, what does the word race mean? Do you have misconceptions about race? You may find answers to these questions and others as you study the basis for dividing a species into subspecies.

IV. THE PROBLEM OF POPULATION

Many authorities view the population problem as the cause of a number of other problems of the world. Is this the case? Can the population problem be solved? Will it be solved? How do we make intelligent decisions about population numbers and other related problems?

V. DRUGS

What makes a substance a drug? What parts of the body does it affect—and how? The psychology of drug use is much talked about, but what about the actual goings-on in the body? Do you know the facts? Will these facts have anything to do with your decisions relating to use or misuse of drugs? These and other questions are considered as the biological aspects of drug use are studied.

study guide

cancer

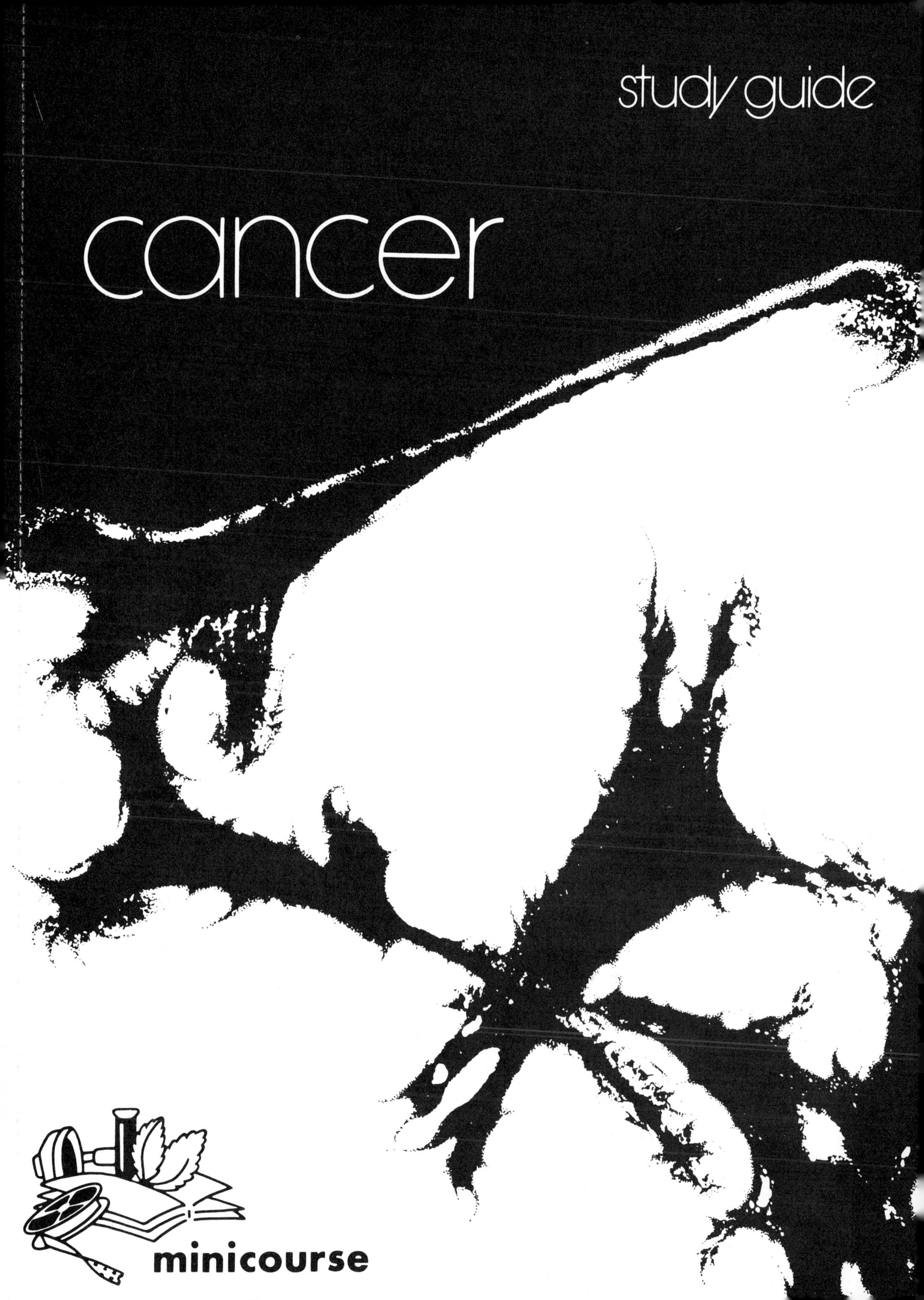

minicourse

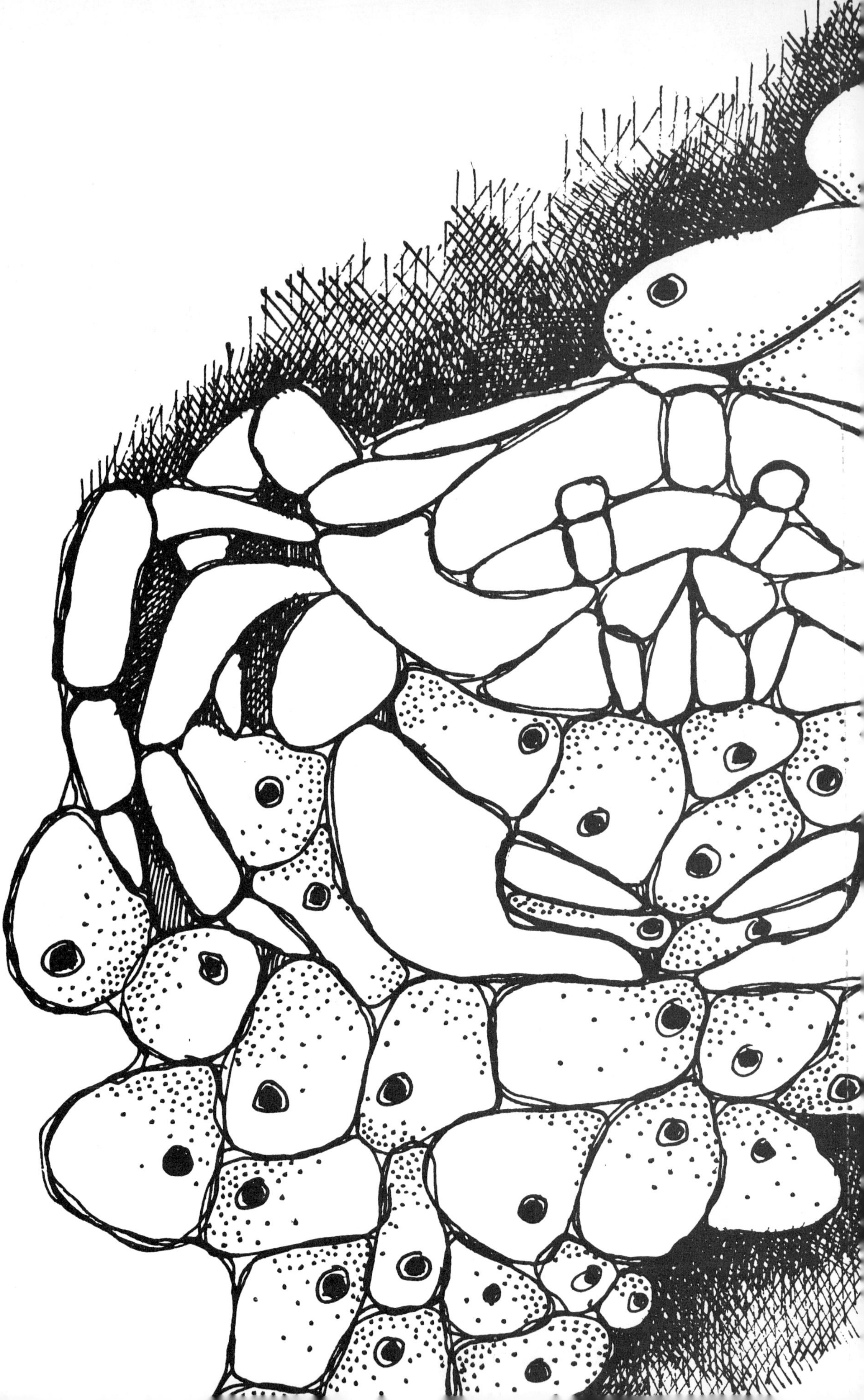

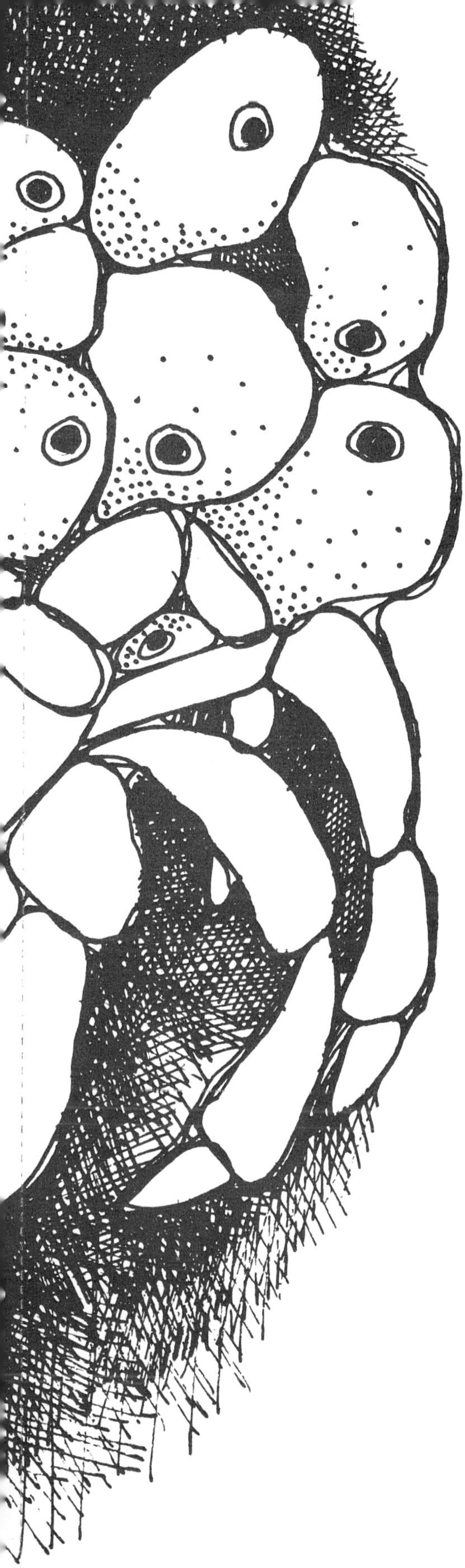

Among the problems facing man today, cancer must certainly rank high. It merits concern by all members of our society because it is no respector of age, sex, or social status. Nearly one out of four persons living today will be treated for cancer at some time in their lives. With the general upsurge in biological research in recent years, a great deal of information has been learned about cancer. What is the nature of the disease? How do cancer cells differ from those that are normal? What causes these differences? What preventions and treatments are available?

PREREQUISITES: A basic knowledge of cell structure and reproduction.

1 Cancer

I. WHAT IS THE NATURE OF CANCER?

EXERCISE 1. What are the similarities and differences between normal and cancerous tissue?

In your study area you will find two small glass jars, one containing a piece of normal lung tissue and the other a piece of lung tissue from an individual who had lung cancer. Both specimens are preserved in 10 percent formalin.

(1) Use forceps to carefully lift the specimens out of their respective jars and place them on open Petri dishes lined with paper toweling. Then examine the specimens macroscopically and with a hand lens (magnifier).

(2) You may wish to include in your examination a gentle squeezing of the tissue, using forceps. *Do not damage the tissue.* The last overlay in the transparent inlay section of the booklet *Cancer Invades* may also be helpful.

(3) Describe any differences and similarities between the two specimens, using the characteristics listed below in Table 1 as a guide. Enter your observations in the appropriate blanks in Table 1.

(4) When you have finished, return the specimens to their proper containers.

TABLE 1.

Comparison of normal and cancerous tissue.

CHARACTERISTIC	NORMAL TISSUE	CANCEROUS TISSUE
Texture (squeeze gently with forceps)		
Color		
Tissue uniformity		
Other		

Cancer: The general term for a variety of disorders characterized by the uncontrolled growth of relatively undifferentiated cells.

In the course of development, individual organs and the whole animal reach a maximum size and then stop growing. You are familiar with the fact that growth stops; there are control mechanisms that limit the size of individual organs and the whole animal. If we are wounded, as by a cut in the skin, the wound starts healing and the new tissue grows to a certain size and then stops growing. Again here we have control mechanisms that limit the growth of the tissue. If one of our appendages, such as a finger, is amputated, it does not grow back, but certain tissues do regenerate if portions of them are removed. An example of this is the liver. Under normal conditions, the liver maintains its size and there is little cellular multiplication. However, if a portion of the liver is cut away, what remains will start regenerating itself until it reaches its original size, and then it stops growing. Thus, after the trauma of removal of a portion of the liver, the control mechanisms that maintain its normal size are altered, and processes are set in motion for the liver to renew the portion that was lost. However, as soon as it reaches its original size, the control mechanisms stop the rapid cell multiplication.

Under some conditions, however, the factors that control the multiplication of cells and limit the size of organs of our body are not operative, and as a result, tissues continue growing. This overgrowth of tissues beyond their normal size or cell number is called cancer.

Dr. Merwin Moskowitz
Department of Biological Sciences
Purdue University

CANCER IS CELLULAR ANARCHY

The term "cancer" means "crab" in Greek. The projecting masses of invading cancer cells in surrounding tissues have been described as resembling the appendages of a crab, the center of growth corresponding to the body.

A. Are All Cancers Alike?

1. general terms
   a. *tumor* (neoplasm)—a mass of cells resulting from uncontrolled growth

(1) *benign tumor*—a cell mass that grows slowly and is surrounded by a capsule that localizes the growth in the tissue in which it originated. A benign tumor is *not* a cancer.

(2) *malignant tumor*—a cell mass that multiplies rapidly and is not restricted by any capsule or membranous sac; it may invade and ultimately destroy surrounding tissues. A malignant tumor *is* a cancer.

b. *metastases*—secondary tumors produced by malignant cells that have migrated from the original site.

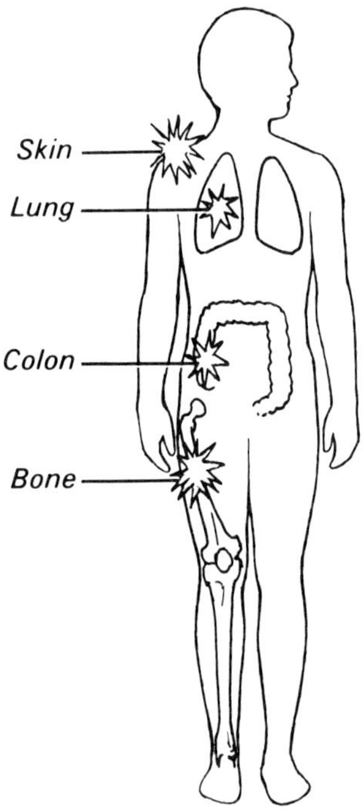

Figure 1. Common cancer sites.

2. classification by origin

   a. *carcinoma*—cancer of epithelial origin. Examples: skin carcinoma; bronchogenic carcinoma (lung cancer arising from a bronchus); carcinoma of a colon (arising from the intestinal mucosa).

   b. *sarcoma*—cancer of mesodermal origin (arising from muscle, bone, cartilage, and connective tissue). Examples: osteogenic sarcoma (bone cancer); leukemia (cancer of the blood-forming organs).

3. classification by consistency of the tumor

   a. solid—example: bronchogenic carcinoma, osteogenic sarcoma

   b. fluid—example: leukemia

B. Normal Cells versus Cancer Cells

1. general structural appearance of individual cells

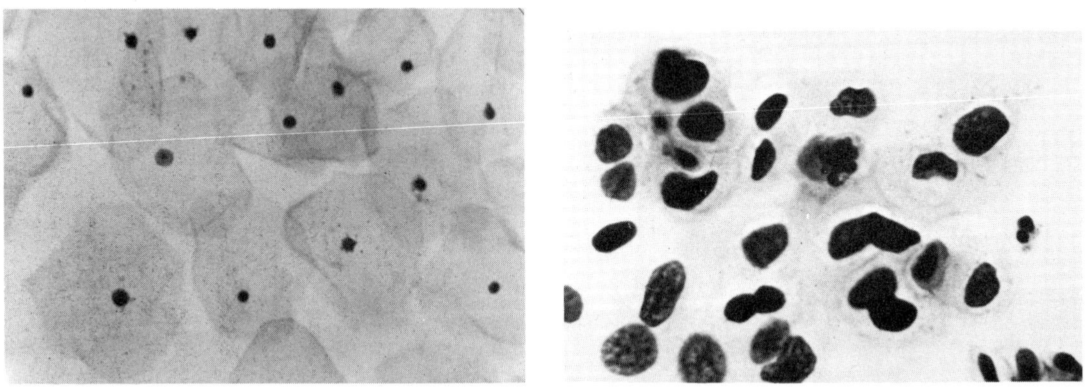

Figure 2. Normal (*left*) and cancerous (*right*) cells from the uterine cervix (550×). (Photos courtesy American Cancer Society, Inc.)

EXERCISE 2. Examination of cancer cells

Reexamine the illustrations in the booklet *Cancer Invades,* noting particularly the differences between normal and cancerous cells of the lungs. Then study Slides 2 to 5. Determine which slides show normal lung tissue and which ones show cancerous lung tissue. Write down the observable differences between the normal and cancerous cells, in the space provided below.

2. orientation

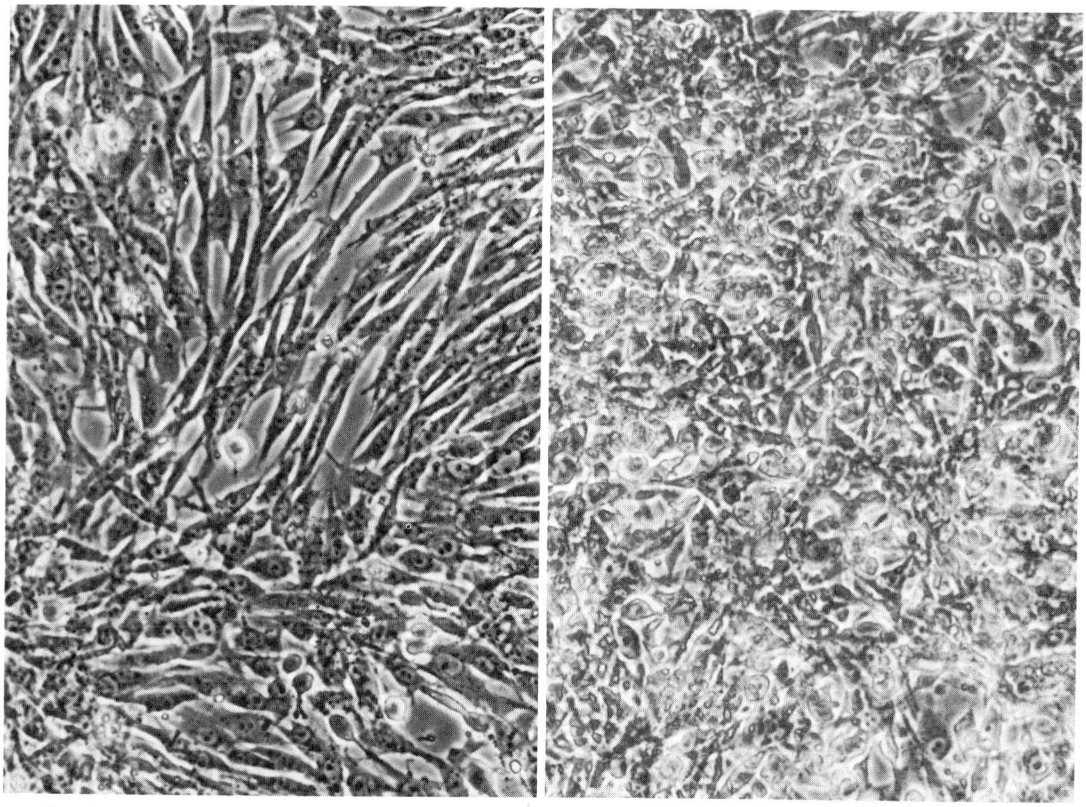

Figure 3. A culture of normal hamster embryo cells (*left*) and cells of similar origin that have become cancerous (*right*). Normal cells are generally single-layered with parallel orientation. Cancer cells lack orderly orientation. (From Sussman, 1973. Photographs courtesy R. Dulbecco, Imperial Cancer Research Fund Laboratories, London, England.)

3. properties of surface membranes

Figure 4. Contact inhibition is displayed by normal cells but not by cancerous cells.

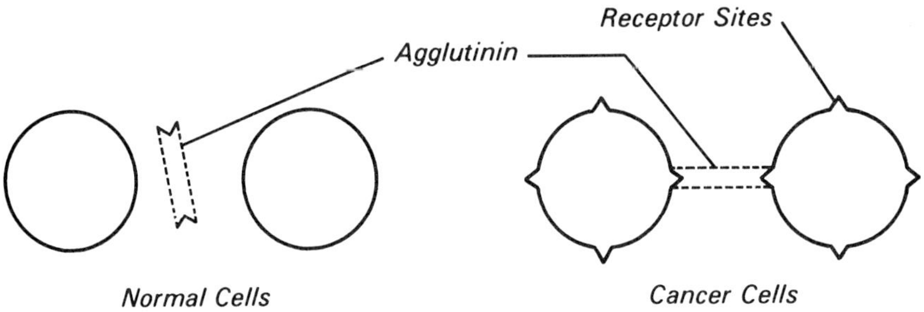

Figure 5. Effects of the addition of plant agglutinins to normal and cancerous cells.

C. Behavior of Cancer Cells

1. *autonomy*—Cancer cells are autonomous, i.e., they have the ability to grow and reproduce without the restraints of normal cell regulatory mechanisms.
2. *metastasis*—Metastasis is the ability of cancer cells to spread both locally and to sites some distance from the original invasion site. Six methods by which cancer may spread are:
   a. by infiltration of tissue spaces
   b. via the lymphatic ducts
   c. via the blood vessels (especially the veins)
   d. along natural body passages such as bronchi and ureters
   e. through natural body cavities such as from the stomach to the ovaries
   f. in experimental animals, by intentional inoculation

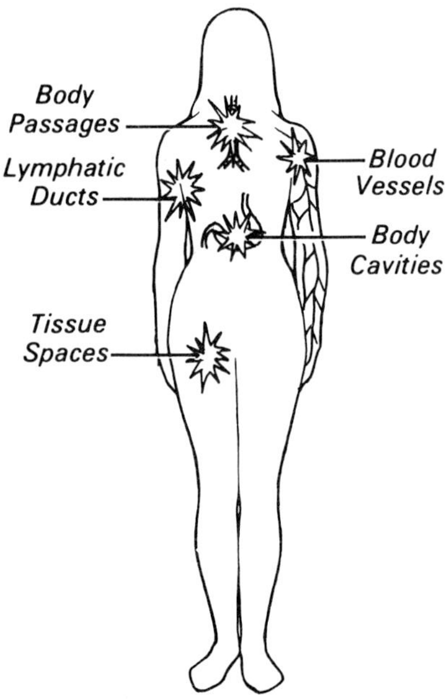

Figure 6. Metastasis of cancer can occur in a variety of ways.

3. *regression*—The reversal of malignant growth into benign or even normal cells occurs occasionally. Regression may be:

   a. spontaneous—The reason the regression occurs is not known. The best example of spontaneous regression in human cancer is probably the regression of neuroblastoma, a cancer of the adrenal gland. This cancer is more often present in the first three years of life than later, and is usually lethal if untreated. However, many reports have demonstrated that partial or complete regression of neuroblastoma has occurred without the help of any induced or therapeutic measure.

   b. induced—Some evidence has been found that regression of cancer may be associated with immune or allergic reactions, acute bacterial infections, endocrine influences, unusual sensitivity, some usually inadequate treatment, interference with the nutrition of cancer cells, surgical removal of the tumor, chemotherapy, or radiation therapy.

II. WHAT IS THE EXTENT OF THE CANCER PROBLEM?

A. Disease Rates—Past and Present

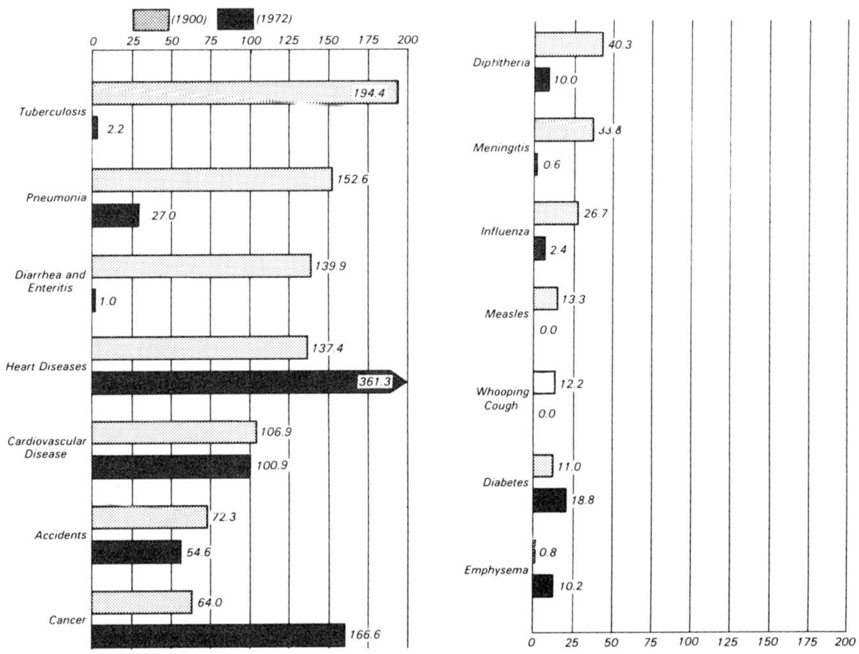

Figure 7. Death rates from selective causes in the United States, 1900 and 1972. (Deaths per 100,000 population.) (Data from the Department of Health, Education, and Welfare, Public Health Service, Annual Report, *Vital Statistics of the United States.*)

B. What Body Sites Are Most Commonly Attacked by Cancer?

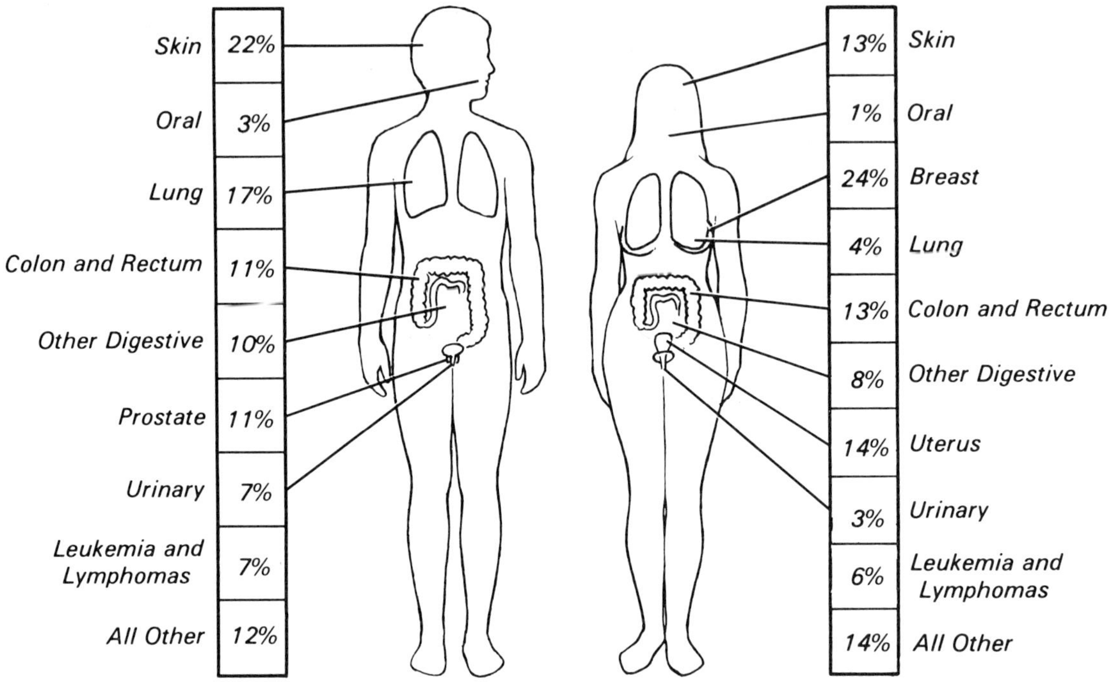

Figure 8. Incidence of cancer by sex and site, 1970. (Courtesy American Cancer Society, Inc.)

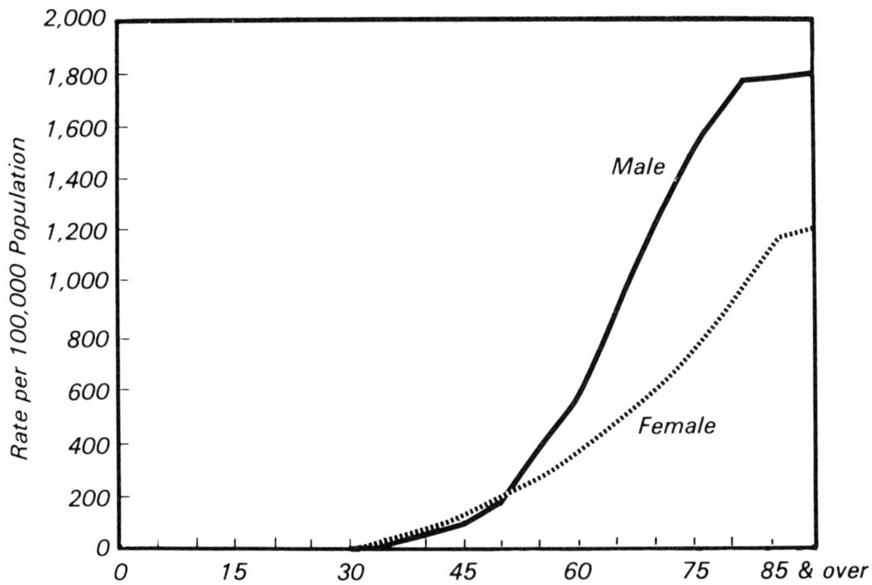

Figure 9. Total cancer death rates by age and sex in the United States, 1972. (Courtesy American Cancer Society, Inc. Data from *Vital Statistics of the United States.*)

EXERCISE 3. Prevalence of cancer

Figures 8 and 9 present recent data regarding the incidence of cancer in the United States. Use this information to answer Questions 1 to 3 below. Then go on to Question 4 and make the prediction requested.

(1) Considering males and females together, what part of the body is most likely to be attacked by cancer?

(2) Among males, what is the most prevalent form of cancer? Among females?

(3) The chances of contracting cancer are about the same for men as for women until about what age?

(4) Predict what you think your chances are of being successfully treated for cancer if you were to contract it.

(a) About 1 in 2 can be treated successfully

(b) About 1 in 5 can be treated successfully

(c) About 1 in 12 can be treated successfully

(d) In most cases life can only be prolonged temporarily, but the cancer not really treated successfully.

C. How Common and Serious Are the Major Types of Cancer?

TABLE 2.

Cancer: the casualty figures, warning signals, and safeguards.*

	SITE	ESTIMATED NEW CASES 1970	ESTIMATED DEATHS 1970	WARNING SIGNAL (WHEN LASTING LONGER THAN TWO WEEKS SEE YOUR DOCTOR)	SAFEGUARDS	COMMENT
1	Breast	67,000	29,000	Lump or thickening in the breast	Annual checkup; monthly breast self-examination	The leading cause of cancer death in women.
2	Colon and rectum	73,000	46,000	Change in bowel habits; bleeding	Annual checkup, including proctoscopy	Considered a highly curable disease when found early during routine physical examinations.
3	Lung	68,000	62,000	Persistent cough, or lingering respiratory ailment	Prevention: heed facts about smoking; annual checkup; chest X-ray	The leading cause of cancer death among men, this form of cancer is largely preventable.
4	Oral (including pharynx)	14,000	7,000	Sore that does not heal; difficulty in swallowing	Annual checkup	Many more lives should be saved because the mouth is easily accessible to visual examination by physicians and dentists.
5	Skin	112,000	5,000	Sore that does not heal, or change in wart or mole	Annual checkup; avoidance of overexposure to sun	Readily detected by observation, and diagnosed by simple biopsy.
6	Uterus	42,000	13,000	Unusual bleeding or discharge	Annual checkup, including pelvic examination and Papanicolaou smear	Mortality has declined 50% during the last 25 years. With wider application of the "Pap" smear, many thousands more lives can be saved.
7	Kidney and bladder	32,000	15,000	Urinary difficulty; bleeding, in which case consult your doctor at once	Annual checkup with urinalysis	Protective measures for workers in high-risk industries are helping to eliminate one of the important causes of these cancers.
8	Larynx	7,000	3,000	Hoarseness; difficulty in swallowing	Annual checkup, including mirror laryngoscopy	Readily curable if caught early.
9	Prostate	35,000	17,000	Urinary difficulty	Annual checkup, including palpation	Occuring mainly in men over 60, the disease can be detected by palpation and urinalysis at annual checkup.
10	Stomach	17,000	16,000	Indigestion	Annual checkup	A 40% decline in mortality in 20 years, for reasons yet unknown.
11	Leukemia	19,000	15,000	Leukemia is a cancer of blood-forming tissues and is characterized by the abnormal production of immature white blood cells. Acute leukemia strikes mainly children and is treated by drugs that have extended life from a few months to as much as three years. Chronic leukemia strikes usually after age 25 and progresses less rapidly.		
				Cancer experts believe that if drugs or vaccines are found that can cure or prevent any cancers they will be successful first for leukemia and the lymphomas.		
12	Lymphomas	23,000	18,000	These diseases arise in the lymph system and include Hodgkin's disease. Some patients with lymphatic cancers can lead normal lives for many years.		

*Data courtesy American Cancer Society, Inc.

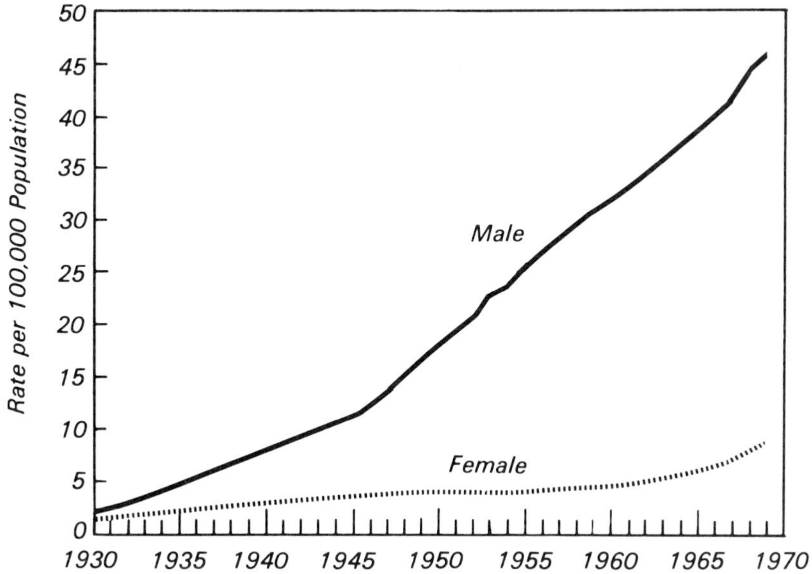

Figure 10. Cancer of the lung death rates by sex in the United States, 1930–1970. Standardized for age on the 1940 population of the United States. (Courtesy American Cancer Society, Inc. Data from *Vital Statistics of the United States.*)

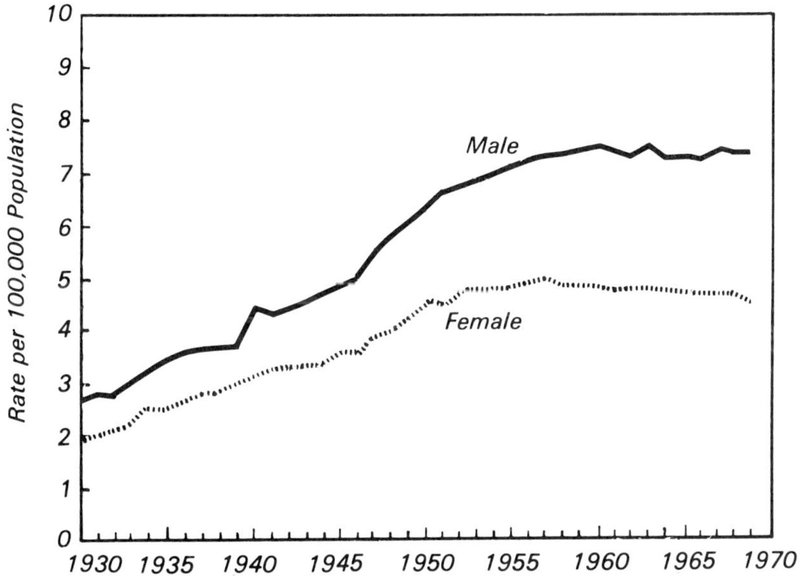

Figure 11. Leukemia death rates by sex in the United States, 1930–1970. Standardized for age on the 1940 population of the United States. (Courtesy American Cancer Society, Inc. Data from *Vital Statistics of the United States.*)

III. WHAT ARE THE CAUSES OF CANCER?

A. Factors Related to the Occurrence of Cancer

1. a specific example of conclusions based on research

Salivary gland tumors were found to be increased more than fivefold among survivors of the atomic bombs who had been exposed to high doses of radiation compared with nonirradiated controls. Despite a small number of proven cases (22 in all), this finding was determined to be highly significant for the entire group and for histologically malignant types, but not for benign tumors.

FROM BELSKEY, ET AL. 1972.

2. higher rates of lung cancer among smokers, liver cancer among South Africans, stomach cancer among Japanese and Scandinavians, mouth cancer among inhabitants of India, skin cancer among Caucasians, cancer among overweight people, leukemia among people exposed to above normal radiation levels; lower rates of breast cancer among Japanese, cervical cancer among Jews, cancer of the penis among circumsized males

B. Radiation

1. ionizing—X ray, radium, nuclear devices
2. nonionizing—ultraviolet rays of the sun

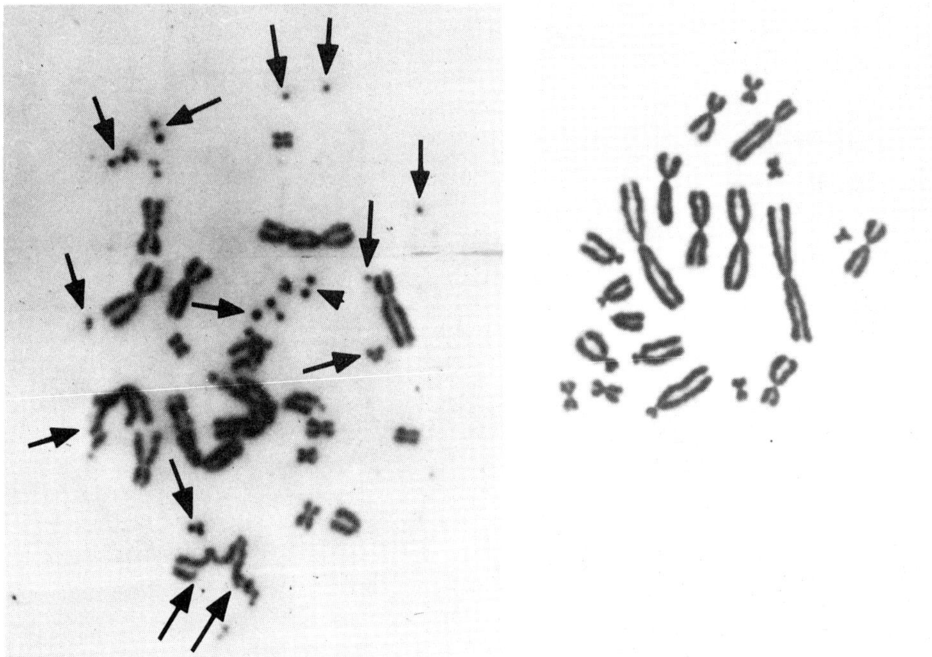

Figure 12. Hamster chromosomes exposed to radiation (*left*) and not exposed (*right*). (From Prescott. 1973.)

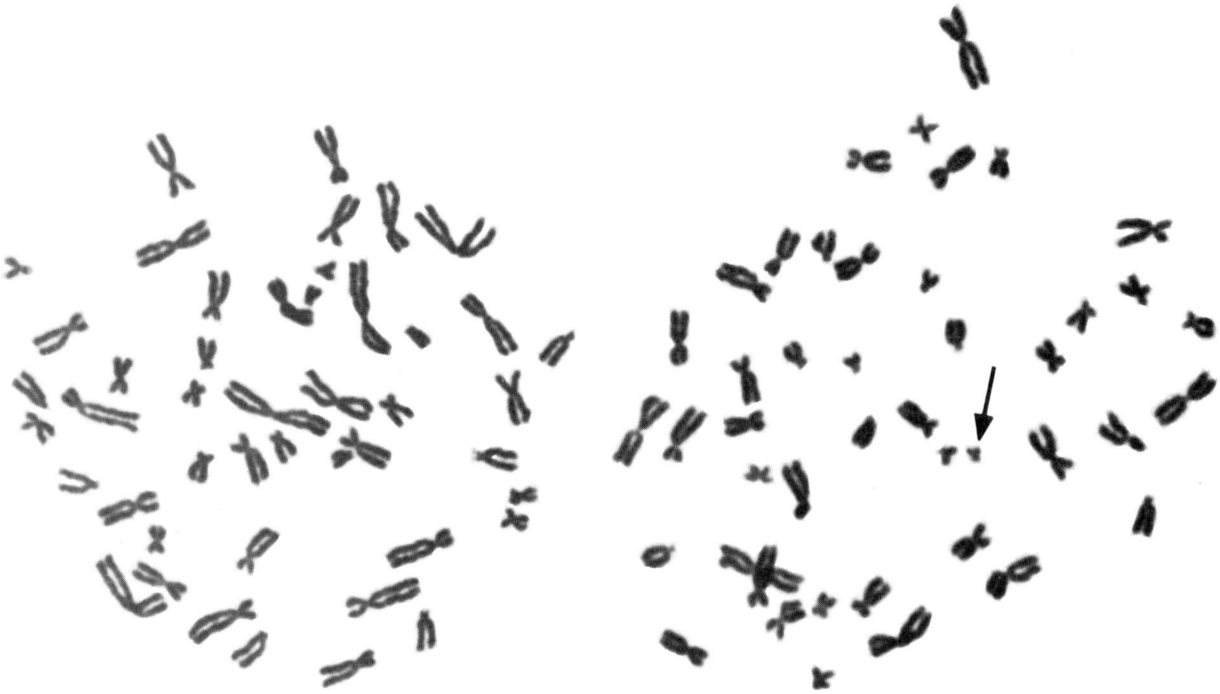

Figure 13. Normal human chromosomes (*left*) and those from a white blood cell of a leukemia patient (*right*). (From Prescott. 1973. Photographs courtesy Dr. Arthur Robinson.)

C. Chemical Agents

TABLE 3.

A short, partial list of chemicals that can cause cancer (chemical carcinogens).*

CHEMICAL CARCINOGENS	
2-acetylaminofluorene	epoxides
aflatoxin B_1	ethionine
aminobiphenyl	ethyl carbamate
anthracene	lead
arsenic	naphthylamine
asbestos	nickel compounds
azo dyes	nitrosamines
benzidine	oil of sassafras
beryllium	oils
cadmium	petroleum products
carbon tetrachloride	pitch
chromium compounds	polycyclic aromatic hydrocarbons
cigarette smoke	pyrrolizidine alkaloids
cobalt	soots
creosote	tars
cycasin (from cycad nuts)	

*From Prescott. 1973.

D. Heredity

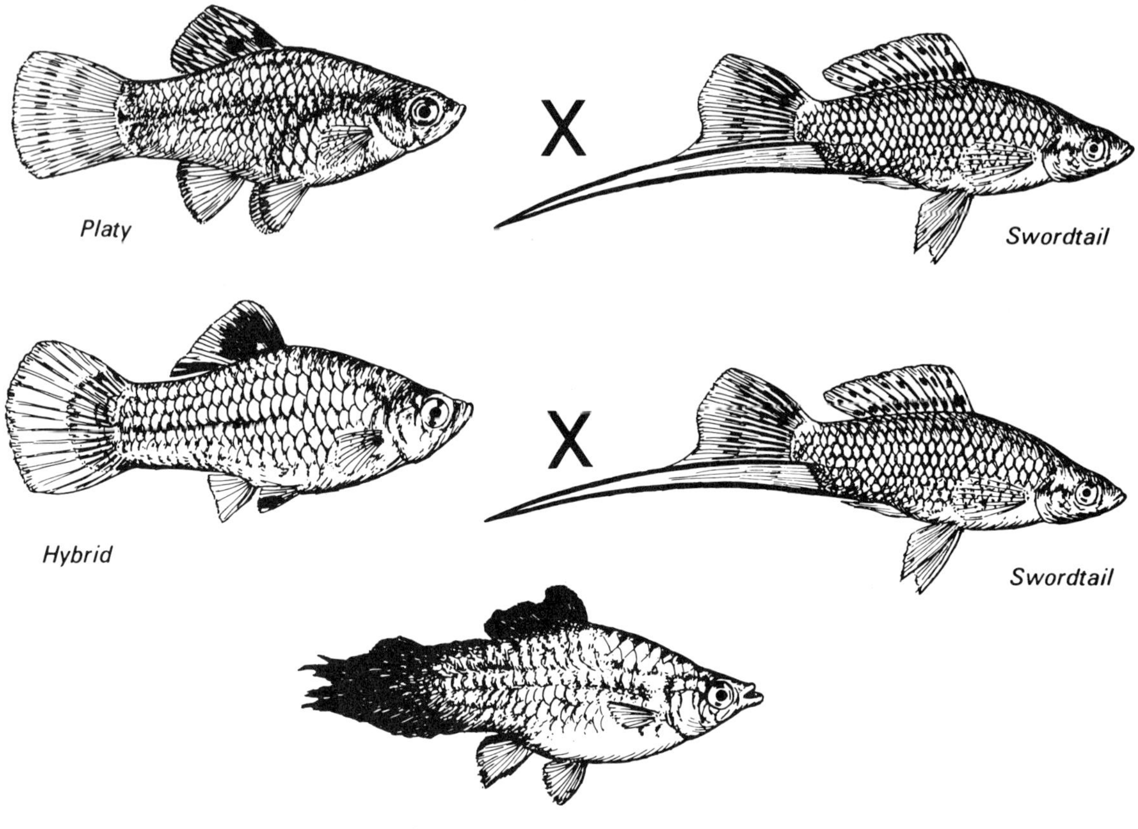

Figure 14. Cancer as a result of crossing platy and swordtail fish. A platy-swordtail cross (*top*) produces a hybrid with pigmented dorsal fin. Crossing this hybrid with another swordtail produces a second generation hybrid that develops melanomas. (Adapted from Prescott. 1973. Drawings based on photographs courtesy Dr. F. Anders.)

E. Viruses

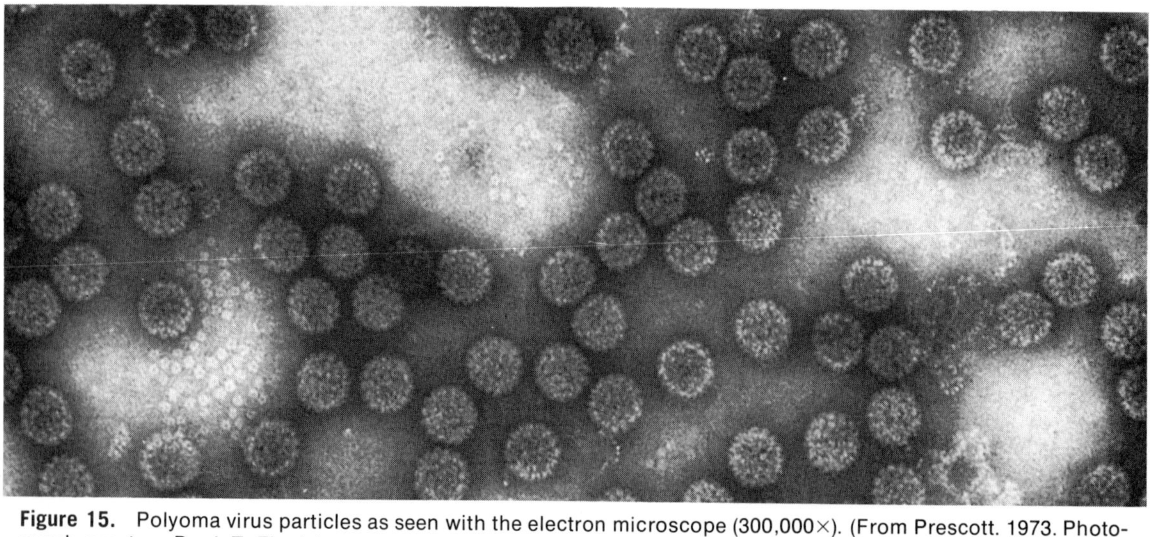

Figure 15. Polyoma virus particles as seen with the electron microscope (300,000×). (From Prescott. 1973. Photograph courtesy Dr. J. T. Finch.)

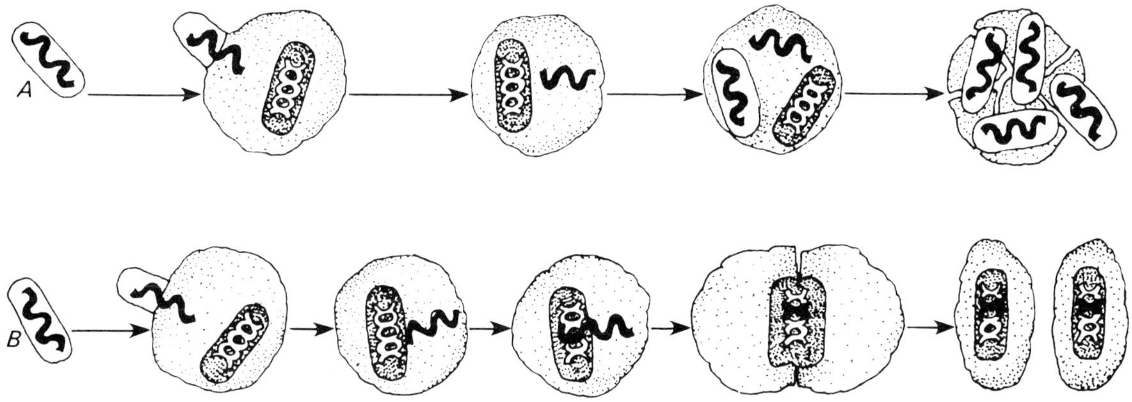

Figure 16. Normal viral infection (*A*) compared to proposed cancer virus transformation (*B*). (Adapted from CRM Books, *Life and Health,* Copyright © 1972 by Ziff-Davis Publishing Company.)

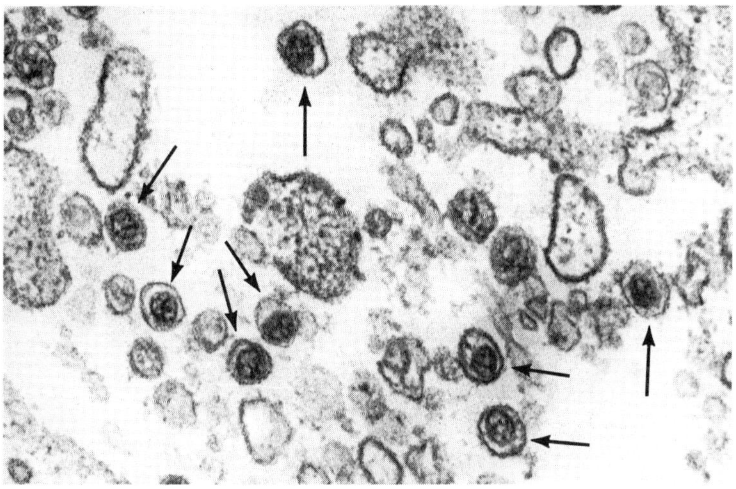

Figure 17. Arrows indicate virus-like particles from a person with leukemia (60,000×). Do they cause the disease? (From Prescott. 1973. Photograph courtesy Dr. Leon Dmochowski.)

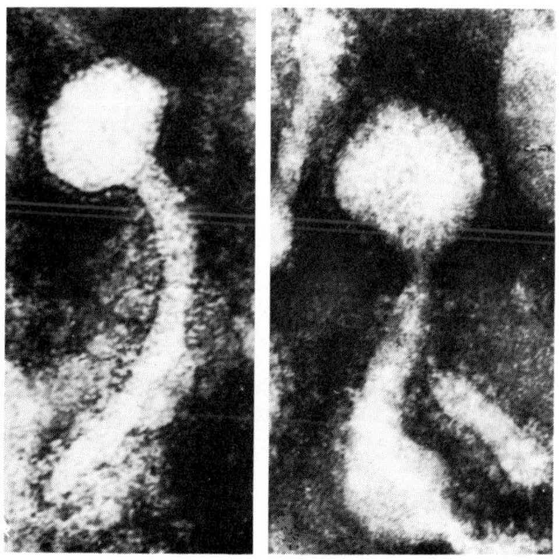

Figure 18. On the left, a virus from mouse milk. On the right, a particle suspected to be a virus from human milk. (150,000×). (From Prescott. 1973. Photographs courtesy Dr. Dan H. Moore.)

F. Immune Theory of Cancer

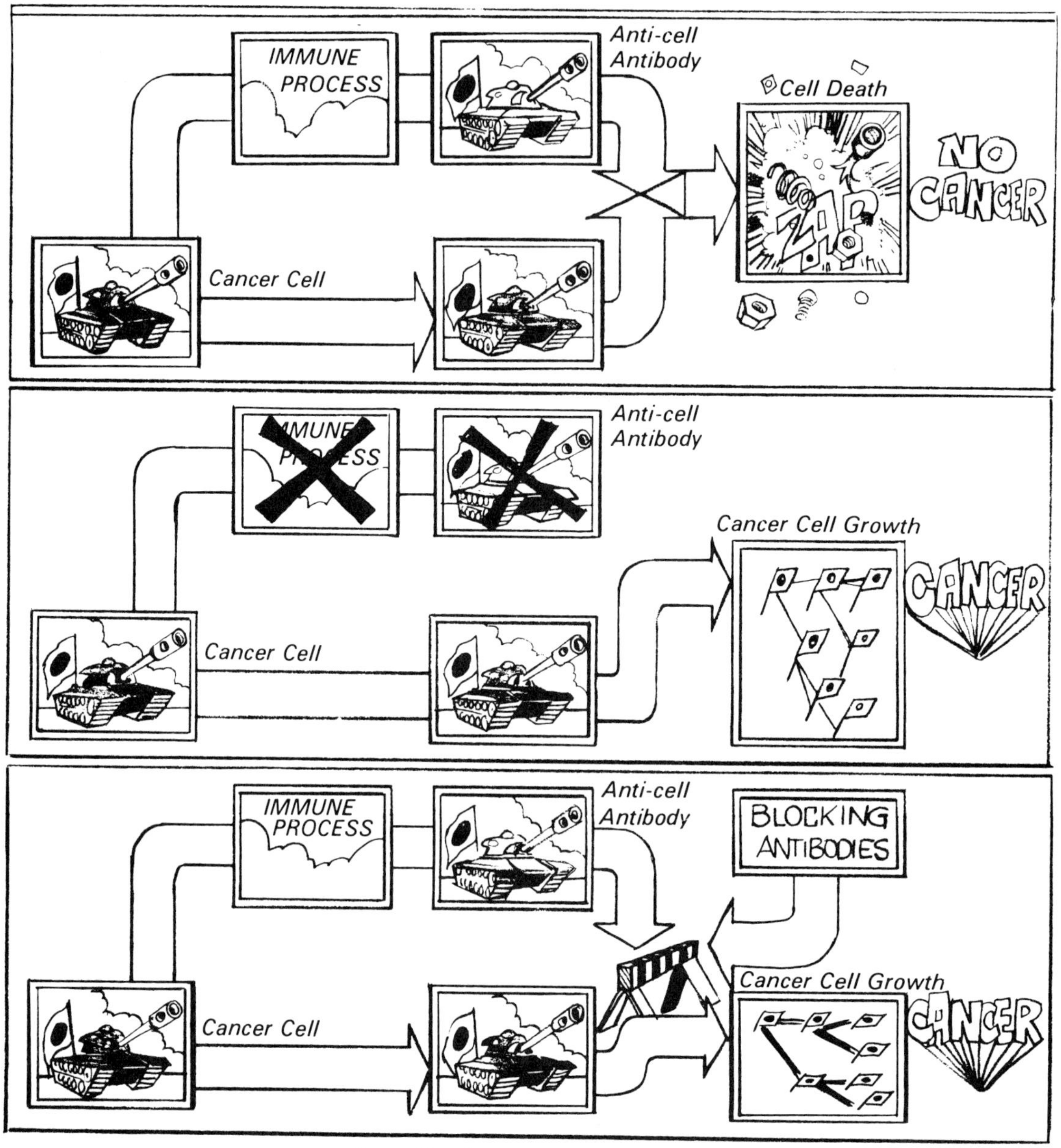

Figure 19. A proposed cancer theory involving the body's immune mechanisms. (Redrawn from an illustration by Ron Estrine, from CRM Books, *Life and Health,* Copyright © 1972 by Ziff-Davis Publishing Company.)

G. Factors Believed to Be Responsible for Cancer – A Summary

1. radiation
2. chemical agents
3. heredity
4. viruses
5. the immune system

IV. WHAT CAN BE DONE TO PREVENT OR TREAT CANCER?

A. Avoidance of Known Carcinogens

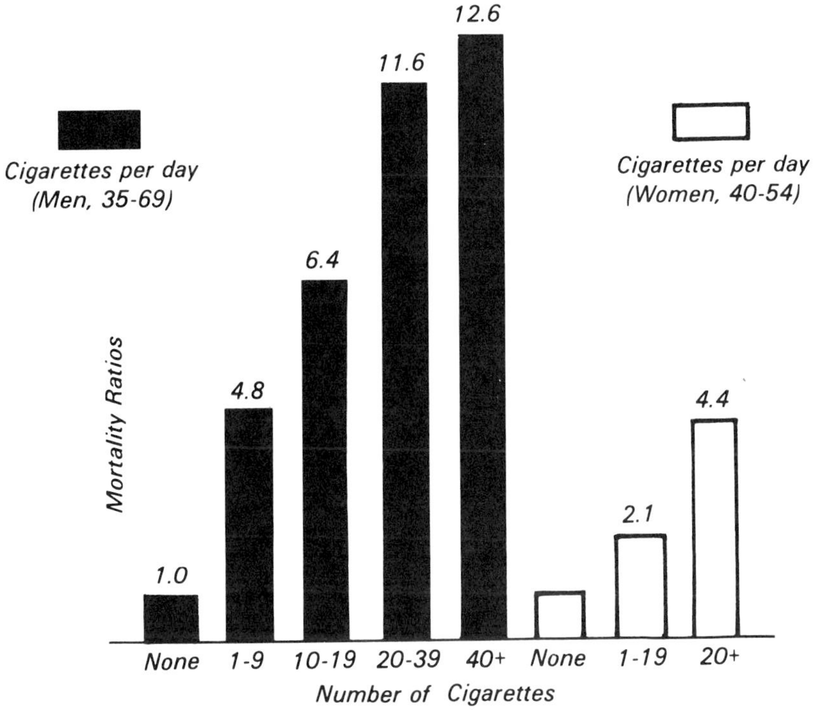

Figure 20. Relationship of lung cancer death rates to number of cigarettes smoked per day. (From *Healthful Living* by H. S. Diehl and W. Dalrymple. Copyright © 1968 by McGraw-Hill Book Company. Used with permission of McGraw-Hill Book Company.)

Figure 21. Exposure to a known carcinogen.

B. Cancer Treatment

1. surgical removal of cancer cells

TABLE 4.

Typical stages in the development of lung cancer.

(a)	Lung cancer is present for several months before it is visible on a chest X ray.
(b)	Once visible, it is still almost twenty months earlier than noticeable physical symptoms.
(c)	Following the appearance of symptoms, the patient typically delays for about three months before seeking medical attention.
(d)	Lung cancer tends to mimic other respiratory diseases. Consequently, doctors may spend four months trying to clear up symptoms with antibiotics.

2. X rays and other radiation

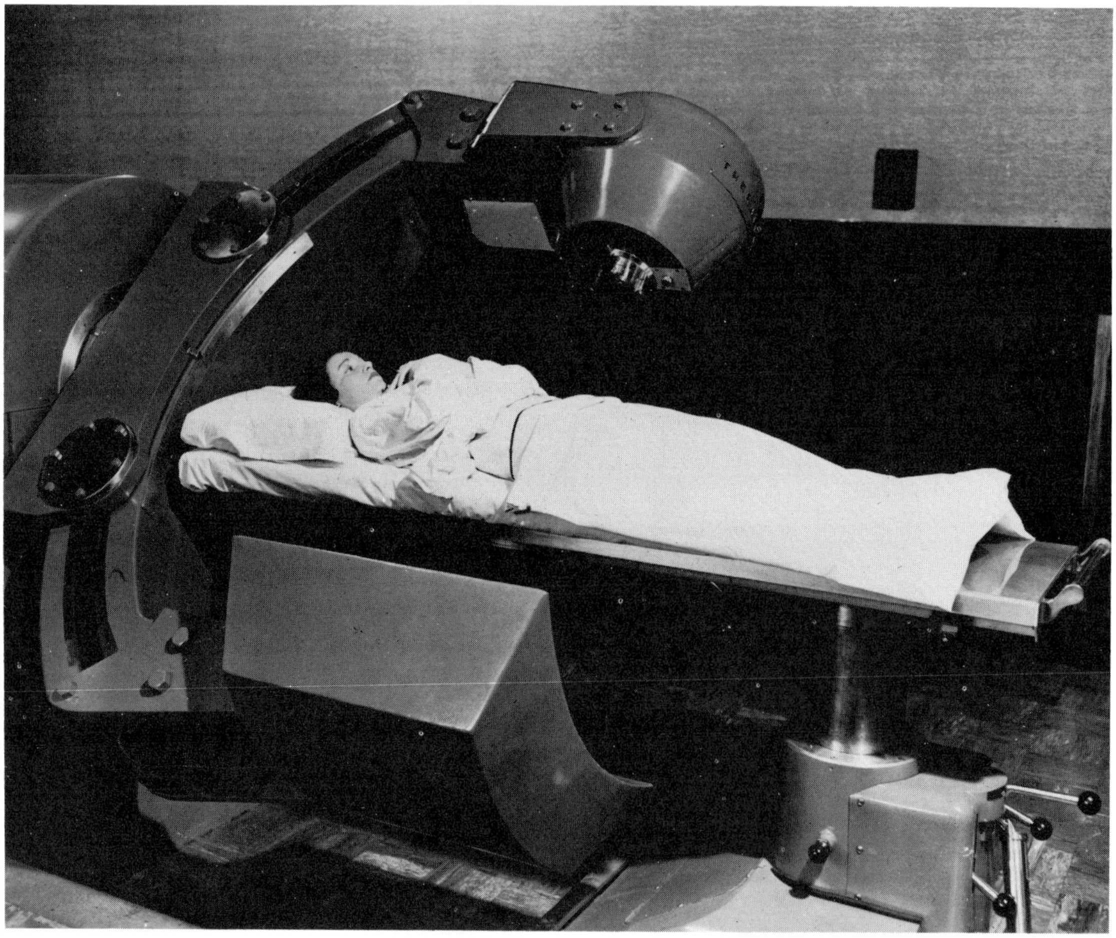

Figure 22. A woman receiving cobalt radiation cancer treatment. (Photograph courtesy American Cancer Society, Inc.)

3. chemotherapy

C. Early Detection

1. examination of the breasts

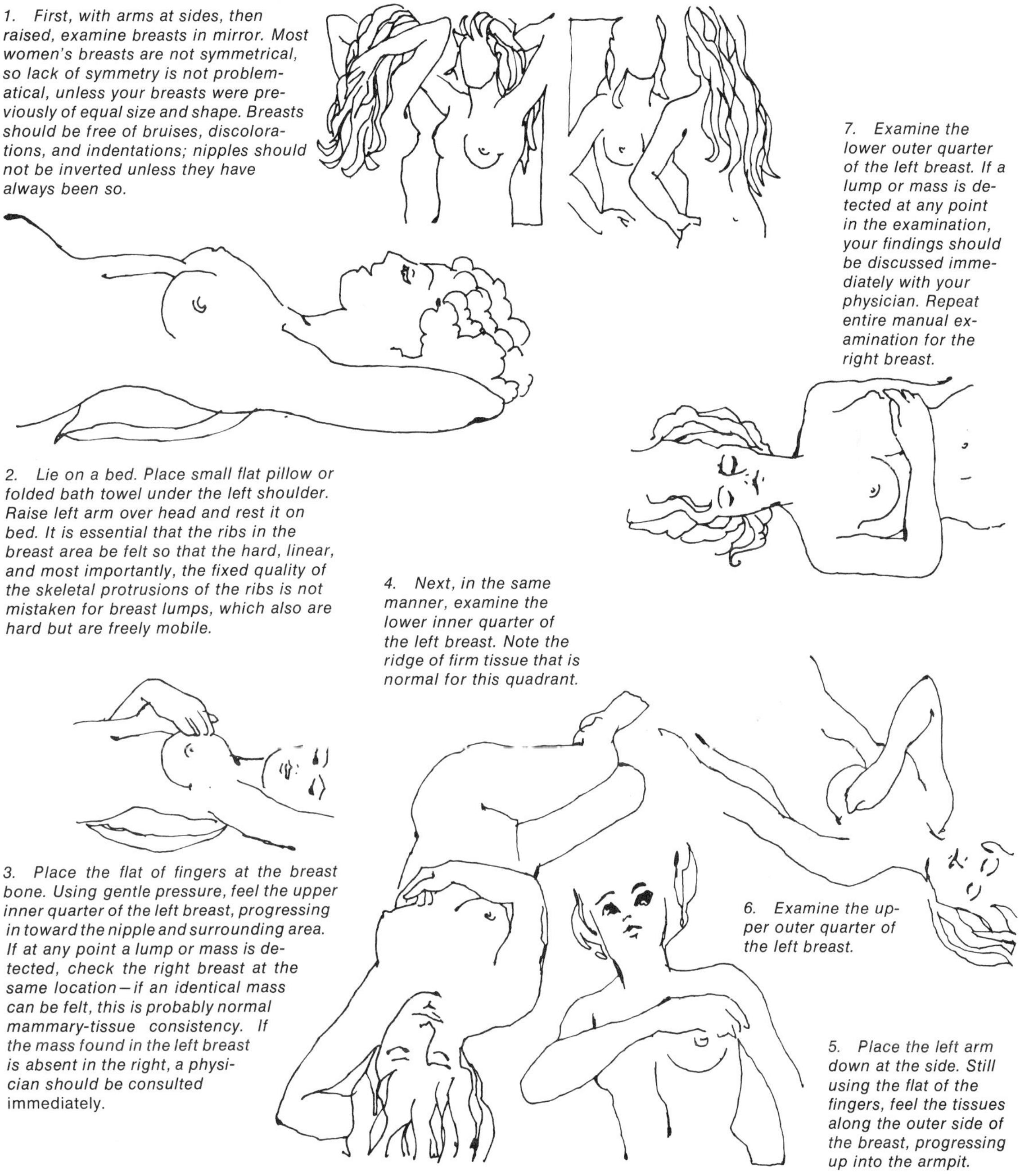

Figure 23. Self-examination of the breasts. (Illustration by Tom Lewis, from CRM Books, *Life and Health*, Copyright © 1972 by Ziff-Davis Publishing Company.)

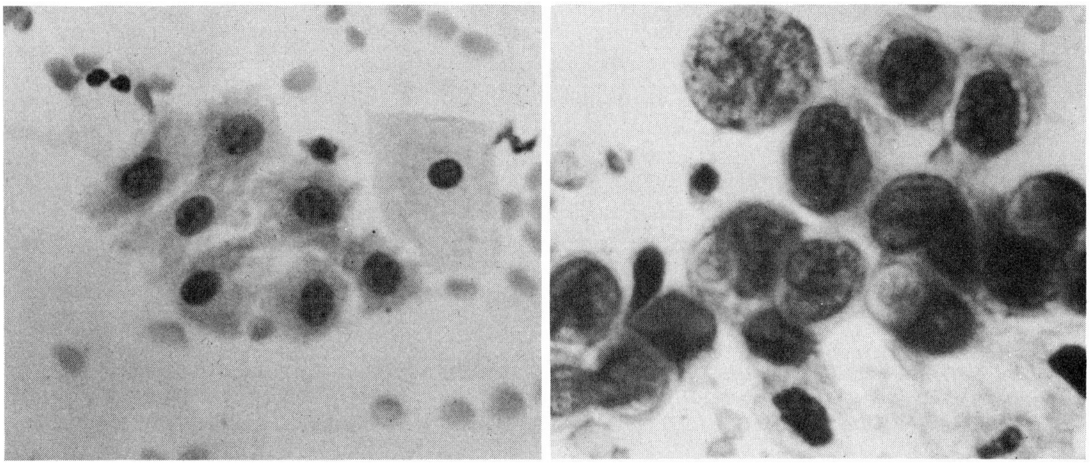

Figure 24. Microscopic views of Pap smears to aid in the detection of cancer of the uterine cervix. Normal cervical cells (*left*) and cancerous cervical cells (*right*). (Photographs courtesy American Cancer Society, Inc.)

2. The American Cancer Society's Seven Warning Signals

*Unusual bleeding or discharge

*A lump or thickening in the breast or elsewhere

*A sore that does not heal

*Change in bowel or bladder habits

*Hoarseness or cough

*Indigestion or difficulty in swallowing

*Change in a wart or mole

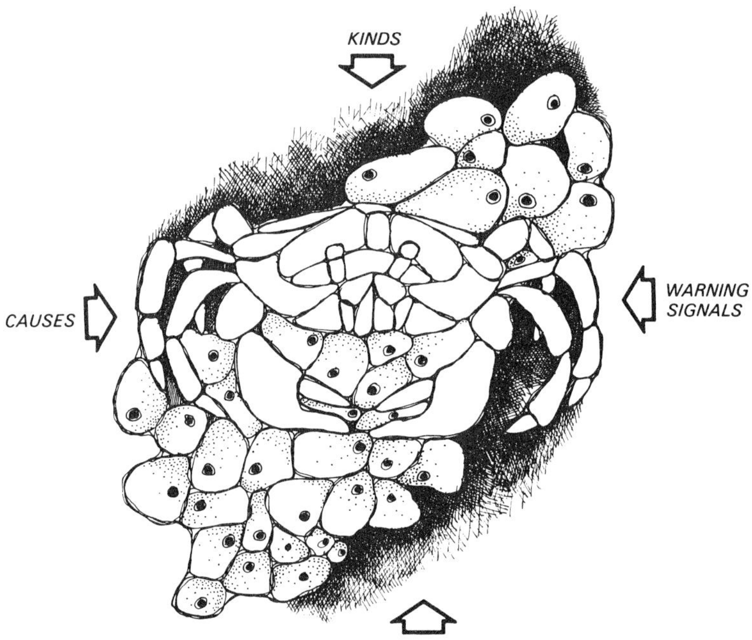

While cancer always has a head start in the race for life, it need not win. Delay, based on fear, often ends in tragedy.

GLOSSARY

AGGLUTINATION: The joining or clumping of suspended particles, cells, or molecules.

ALVEOLI: The tiny air sacs in the lungs through which exchange of gases occurs.

AUTONOMY (OF CANCER CELLS): The ability of cancer cells to grow and reproduce without the restraints of normal cell regulatory mechanisms.

BASAL CELLS: Innermost cells of the deeper epidermis of the skin and respiratory passages.

BENIGN TUMOR: A cell mass that grows slowly and is surrounded by a capsule that localizes the growth in the tissue in which it originated.

BRONCHI (Singular: BRONCHUS): Major air passages leading to the lungs; the branches of the trachea (windpipe).

CANCER: The general term for a variety of disorders characterized by the uncontrolled growth of relatively undifferentiated cells.

CARCINOGEN: Any cancer-producing agent.

CARCINOMA: A cancer of epithelial origin.

CHEMOTHERAPY: The use of chemicals in the treatment and control of disease.

CHROMATIN: The nucleic acid–protein complex of plant and animal cells; the chromosome mass.

CILIA: Short, hairlike processes in certain cells; they are capable of coordinated beating, which produces locomotion or moves particles along the surfaces of the cells.

CONTACT INHIBITION: The inhibition of cell movement upon contact with neighboring cells.

EPITHELIAL CELLS: Cells lining the free surface or tube or cavity of an animal body.

FORMALIN: A chemical widely used in the preservation of biological specimens.

GOBLET CELLS: Mucus-secreting epithelial cells.

INDUCED REGRESSION: A regression in cancerous growth brought on by natural or man-made factors.

LEUKEMIA: Cancer of the blood characterized by an increase of white blood cells.

MALIGNANT TUMOR: A cell mass that multiplies rapidly and is not restricted by any capsule or membranous sac; it may invade and ultimately destroy surrounding tissues. A cancer.

MELANOMA: A cancer resulting from the uncontrolled proliferation of pigment-containing cells, usually in the skin.

METASTASES: Secondary malignant tumors that have migrated from the original site.

MITOSIS: Nuclear division involving replication of each chromosome and distribution of the original number and kinds of chromosomes to the nucleus of each resultant cell.

PAP TEST: A technique for the early diagnosis of uterine cancer, making use of a special staining technique that differentiates diseased tissue; named for its developer, George Papanicolaou.

POLYOMA VIRUS: A virus associated with tumors in certain animals.

REGRESSION: The reversal of malignant growth into benign or even normal cells.

SARCOMA: A cancer of mesodermal origin.

SPONTANEOUS REGRESSION: Regression not attributable to any of the known environmental factors.

TUMOR: Mass of cells resulting from an abnormal growth; also referred to as a neoplasm.

REFERENCES

Arehart-Treichel, J. 1975. Is the great cancer cure already with us? Science News. *107*:26–27.

Beattie, E. J. 1974. Lung cancer. CA-A Cancer Journal for Clinicians. *24*(2):96.

Belsky, J. L., et al. 1972. Salivary gland tumors in atomic bomb survivors, Hiroshima-Nagasaki, 1957–1970. J.A.M.A. *219*(7):864–868.

Case, J. F., and V. E. Stiers. 1971. Biology: observation

and concept. The Macmillan Company, New York. pp. 385–390.

Communications Research Machines Books. 1972. Life and health. Communication Research Machines, Inc., Del Mar, California. 643 pp.

Diehl, H. S., and W. Dalrymple. 1968. Healthful living. Eighth Edition. McGraw-Hill Book Company, New York. pp. 45–64.

Dubos, R., and M. Pines. 1968. Health and disease. Time-Life Books, Inc., New York. pp. 90–93.

Ebert, J. D., et al. 1973. Biology. Holt, Rinehart and Winston, Inc., New York. pp. 246–249.

Gerking, S. D. 1974. Biological systems. Second Edition. W. B. Saunders Company, Philadelphia. pp. 159–160.

Heimler, C. H., and J. D. Lockard. 1969. Focus on life science. Charles E. Merrill Publishing Company, Columbus, Ohio. 561 pp.

Kogan, B. A. 1970. Health: man in a changing environment. Harcourt Brace Jovanovich, Inc., New York. pp. 198–211.

Prescott, D. M. 1973. Cancer: the misguided cell. Pegasus: Bobbs-Merrill Company, Inc., Indianapolis. 186 pp.

Stephens, G. C., and B. B. North. 1974. Biology. John Wiley and Sons, Inc., New York. pp. 436–439.

Sussman, M. 1973. Developmental biology: its cellular and molecular foundations. Prentice-Hall, Inc., Englewood Cliffs, New Jersey. pp. 250–259.

Takahashi, M. 1971. Color atlas of cancer cytology. Georg Thieme Verlag, Stuttgart. 354 pp.

study guide

disease and technology

minicourse

THE CONTROL OF A DISEASE

1. *Initial consultation with the doctor including review of medical history*

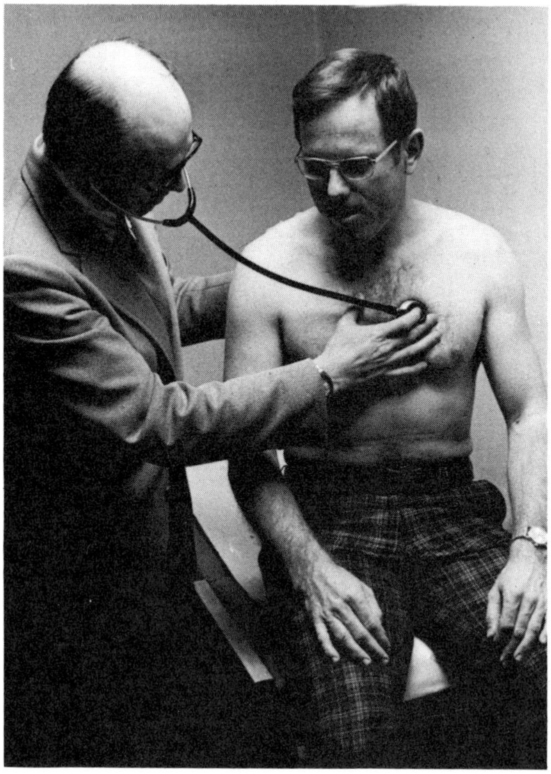

2. *Physical examination*

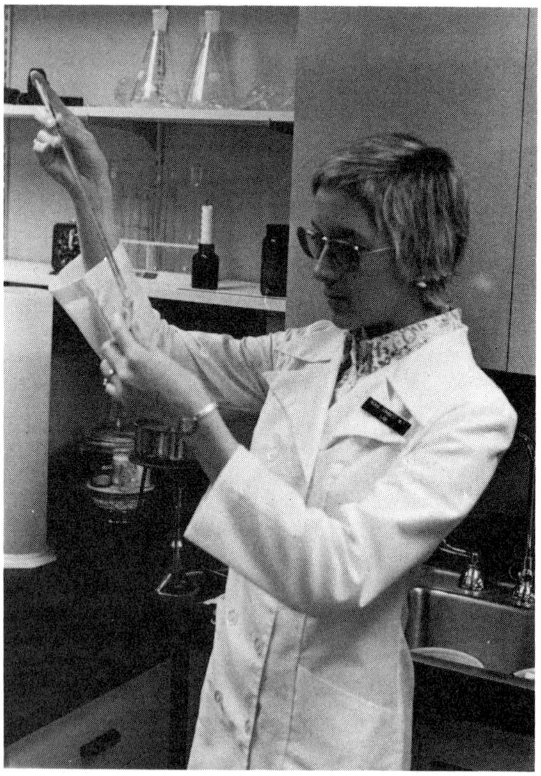

3. *Lab tests and analysis*

PREREQUISITES: There are no prerequisites for this minicourse, although a basic knowledge of simple human genetics will aid in understanding the section pertaining to genetic disease.

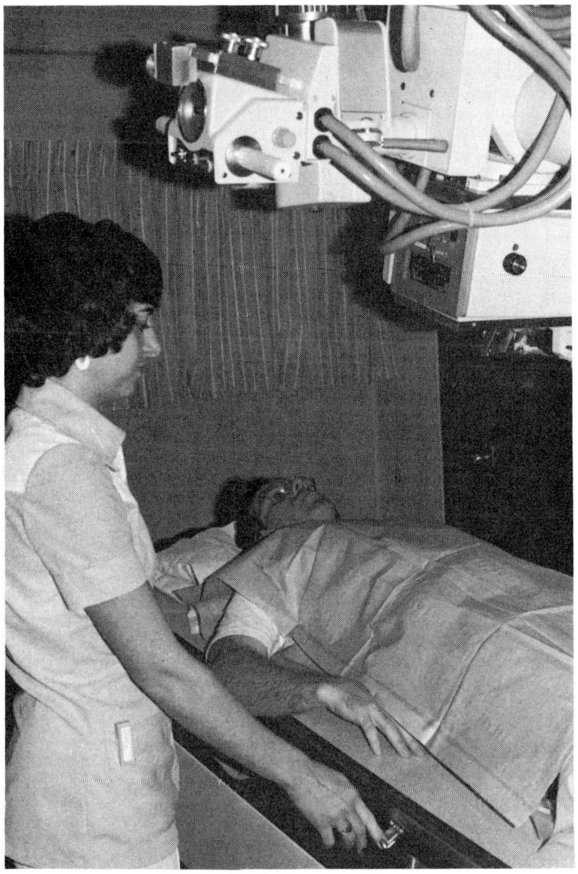

4. X ray examination of colon following barium enema

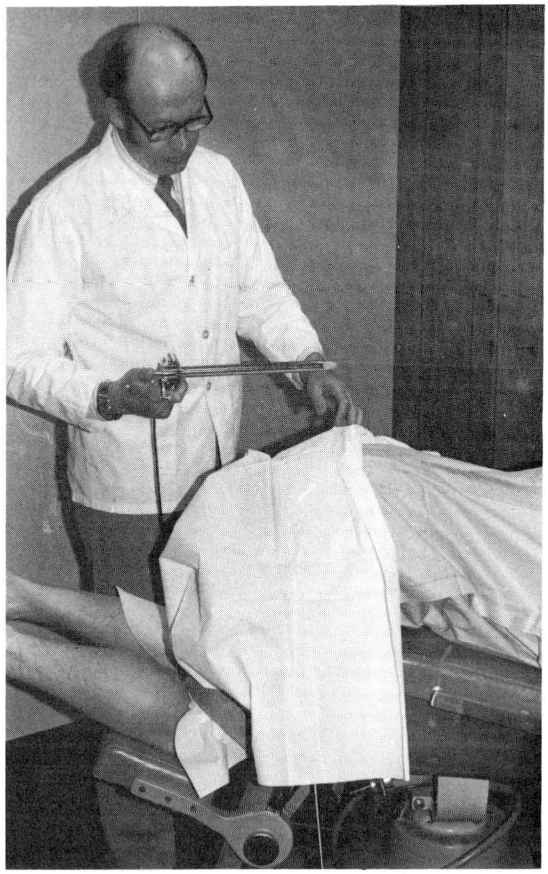

5. Preparation for proctoscopic examination

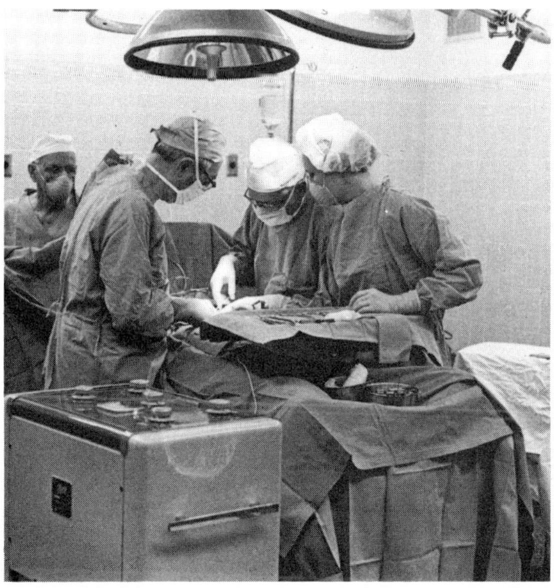

6. Surgical removal of intestinal section containing malignant tissue

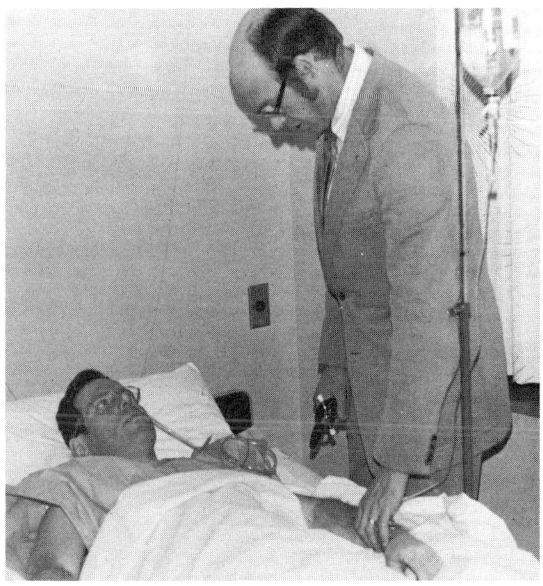

7. Postoperative recovery and evaluation

I. WHAT DOES IT MEAN TO BE HEALTHY? AND HOW DO WE STAY THAT WAY?

A. Symptoms of Disease and Health

EXERCISE 1.

(1) Symptoms of disease:

(2) Symptoms of health:

B. Health Defined by the World Health Organization

Health: A state of complete physical, mental, and social well-being and not merely the absence of disease and infirmity.

Figure 1. Are the patterns of health among these people as variable as the people themselves? (Photo courtesy Monkmeyer Press Photo Service, New York.)

C. Health as a Continuum

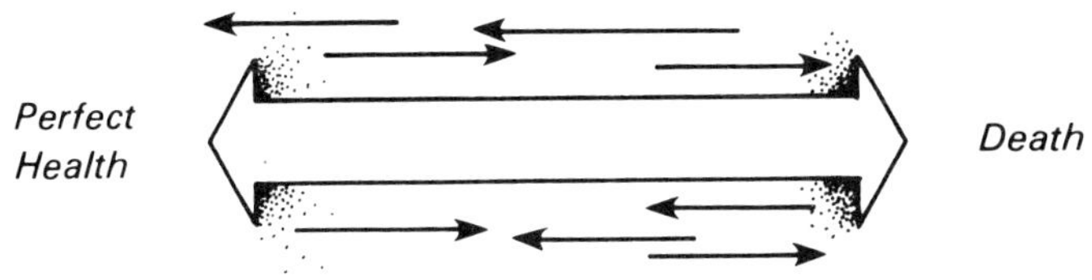

Figure 2. The extremes of health.

II. WHAT IS TECHNOLOGY'S APPROACH TO DISEASE?

A. Determining Cause and Effect

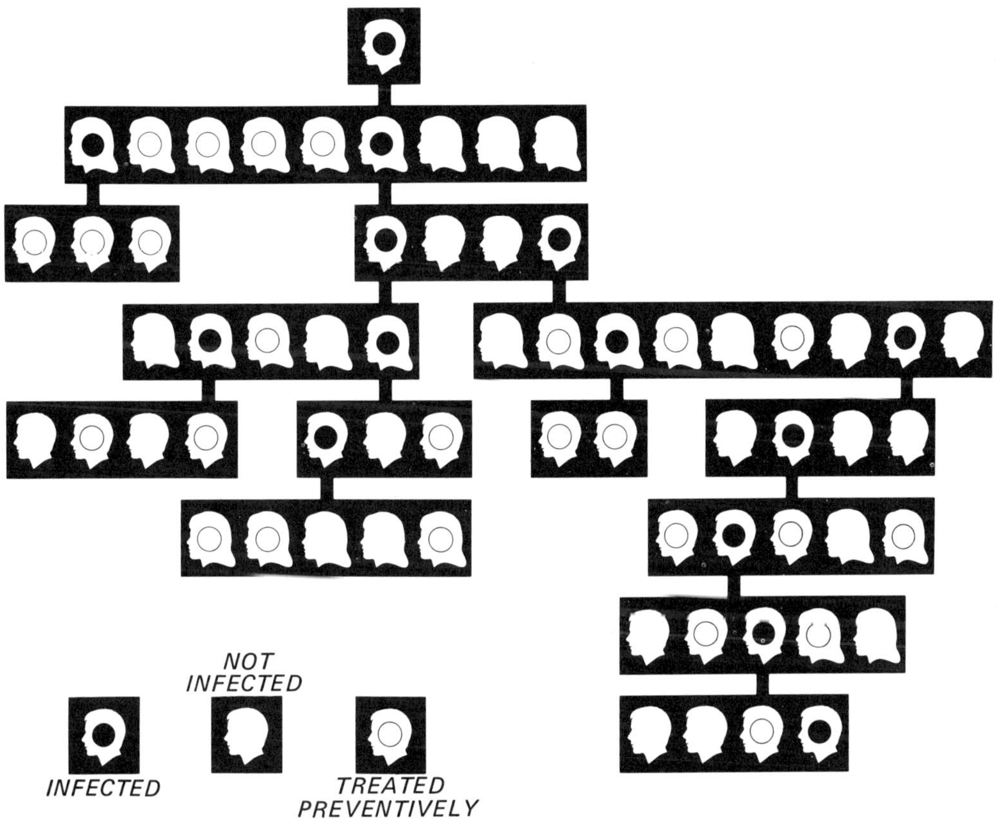

Figure 3. Infectious chain of syphilis at a U.S. high school.

Epidemiology: The science of medicine dealing with the incidence, distribution, and control of disease in a population.

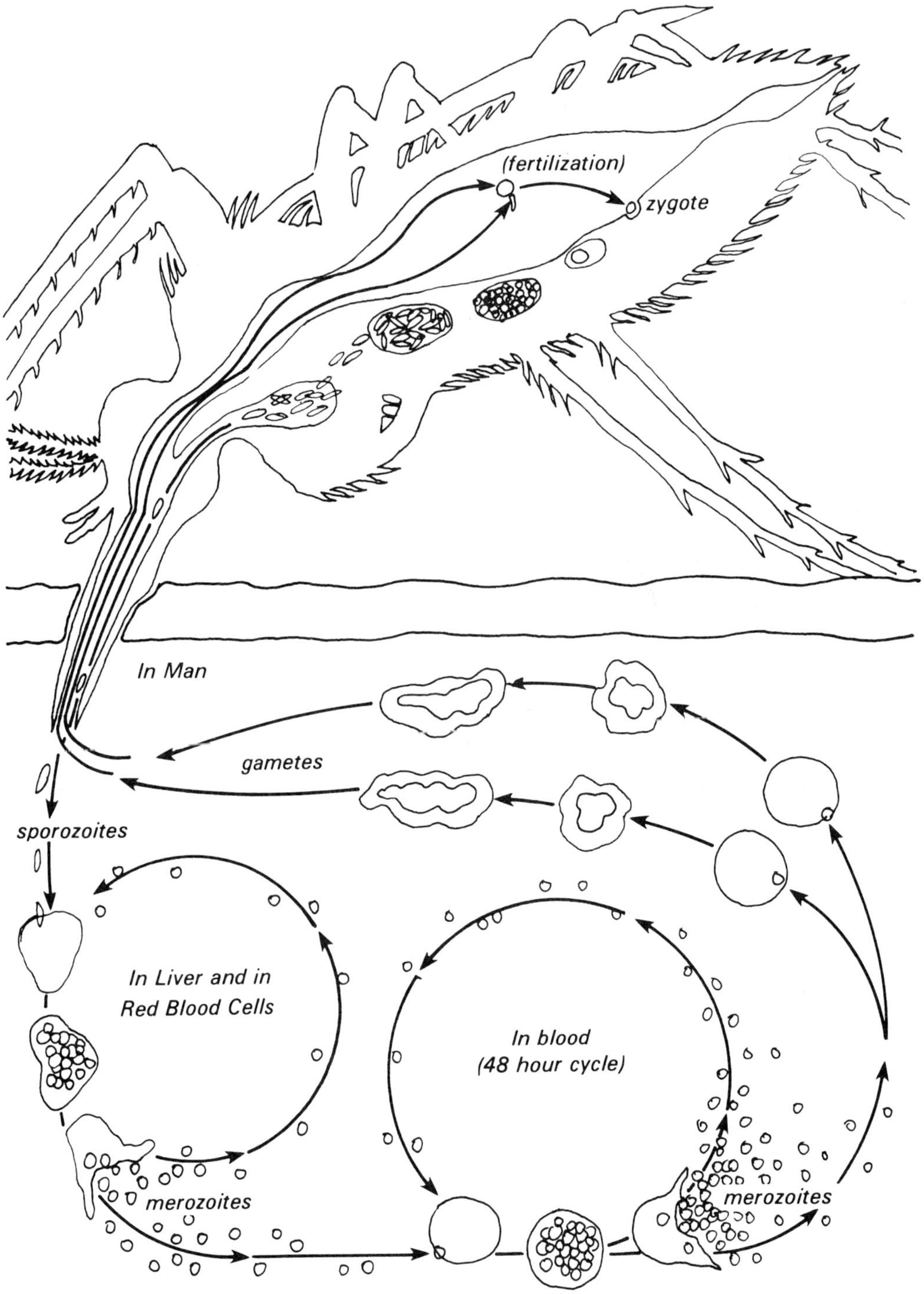

Figure 4. Life cycle of *Plasmodium vivax*, a malarial fever parasite.

TABLE 1.

The classic epidemiological model.

HOST–AGENT–ENVIRONMENT	
INFECTIOUS DISEASE (SMALLPOX)	**NONINFECTIOUS DISEASE (HEART DISEASE)**
1. Environment—one that favors the development and transmission of the virus.	1. Environment—age and behavior leading to development of the disease.
2. Agent—the virus itself, an infectious agent spread by contact.	2. Agent—the inducing factors such as high blood pressure, diet, smoking, lack of exercise, excess weight, and so on.
3. Host—a human being susceptible to the agent (virus)	3. Host—a human being susceptible in terms of genetic predisposition, effects of other diseases, age, and behavior and inducing factors.

B. The Usefulness of Statistics in Understanding and Combating Disease

1. *Mortality*—the number of deaths in a population, usually expressed by cause, per 1000 people in the population
2. *Morbidity*—the occurrence or condition of being diseased, expressed for each disease per 1000 population
3. *Epidemic*—a disease that affects many individuals in an area within a relatively short period of time
4. *Pandemic*—a geographically widespread epidemic
5. *Endemic*—descriptive of a disease common to a certain geographical region at a relatively constant level

III. WHAT ARE THE CHANGING PATTERNS OF DISEASE?

EXERCISE 2. Why have disease patterns changed?

From observations of the data in Figure 5, answer the following questions:

(1) Identify the three leading causes of death in 1900 and 1972.

1900 1972

(2) Suggest two reasons for the change in the pattern of diseases.

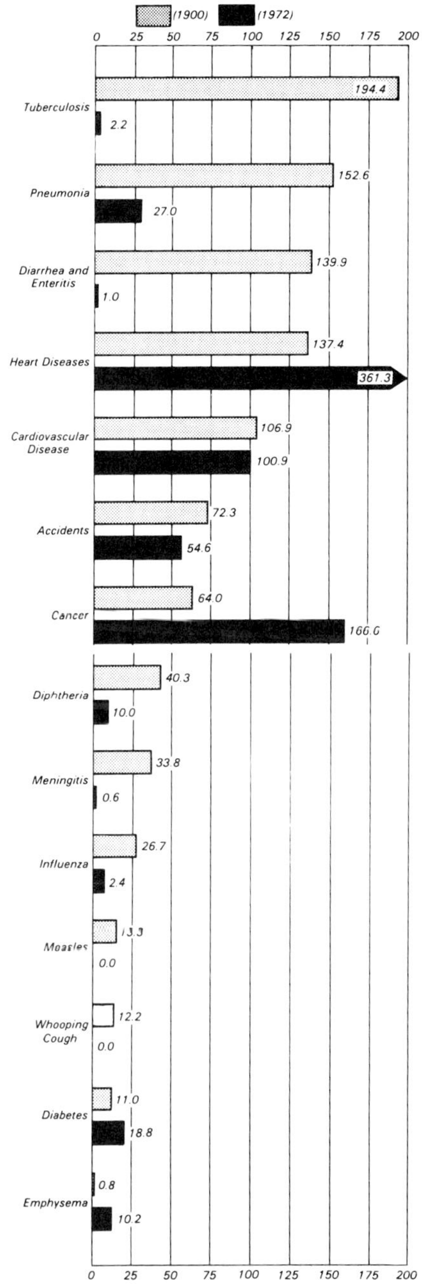

Figure 5. Leading causes of death in the United States, 1900 and 1972. (Data from Department of Health, Education, and Welfare, Public Health Service, Annual Report, *Vital Statistics of the United States.*)

IV. WHAT IS DISEASE AND WHAT ARE ITS CAUSES?

A. Disease Defined:

B. The Various Disease-Causing Factors

1. heredity
2. diet
3. infection
4. environment
5. stress
6. degenerative processes

EXERCISE 3. Diseases and their causes

In Table 2 match the disease-causing factor(s) listed in Part B above with those deaths caused by disease in the left-hand column. Place the name or number of the factor in the right-hand column. Remember that a disease may be attributed to more than one factor.

TABLE 2.

The leading causes of death in the United States in 1972.*

CAUSE OF DEATH	DISEASE-CAUSING FACTOR(S)
1. Disease of the heart	
2. Cancer	
3. Cerebrovascular disease (stroke)	
4. Accidents	
5. Influenza and pneumonia	
6. Diabetes mellitus	
7. Symptoms and ill-defined symptoms	
8. Diseases of infancy	
9. Arteriosclerosis**	
10. Cirrhosis of liver	
11. Suicide	
12. Emphysema	
13. Homicide	
14. Congenital anomalies	
15. Nephritis and nephrosis†	
16. Ulcer of stomach and duodenum	
17. Hypertension	
18. Infections of kidney	
19. Hernia and intestinal obstructions	
20. Bronchitis	
21. Tuberculosis	
22. Other infective and parasitic diseases	

*Data from Department of Health, Education, and Welfare, Public Health Service, Annual Report, *Vital Statistics of the United States.*

**Medical term for hardening of the arteries.

†Nephritis, also called Bright's disease, is the medical term for inflammation of the kidneys; nephrosis is a term for noninflammatory degeneration of the kidneys.

C. A More Detailed Examination of the Causes of Disease

1. What role does heredity play as a cause of disease?

   a. two examples of human genetic problems

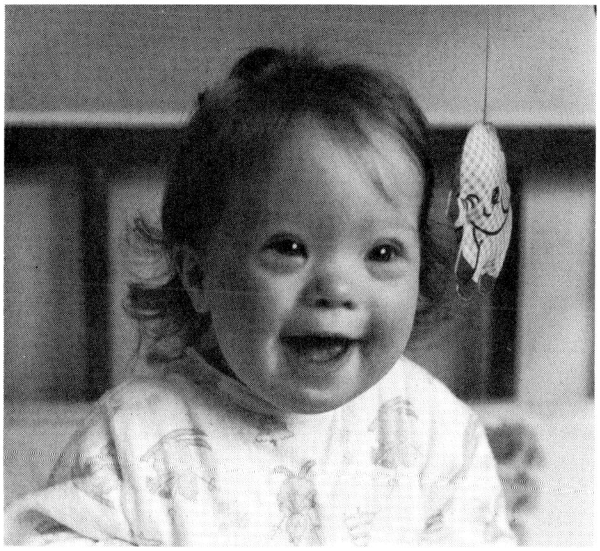

Figure 6. A child with Down's syndrome. (From Smith and Wilson. 1973.)

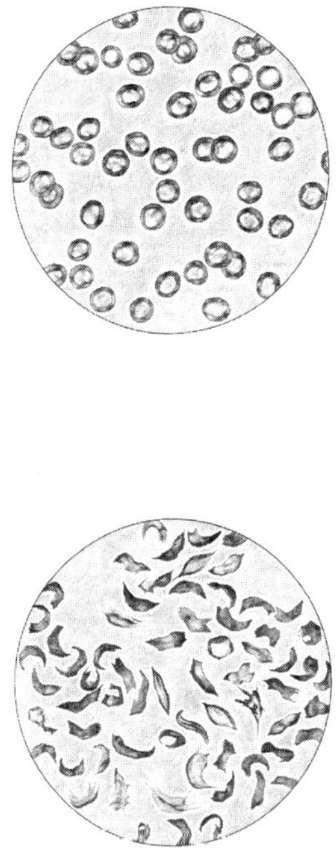

Figure 7. Red blood cells. Upper circle indicates normal cells, the lower circle, typical sickle-shaped cells.

   b. basic types of hereditary problems

   (1) chromosome abnormalities

   (a) Down's syndrome—affected person has 47 rather than the normal 46 chromosomes

   (b) Turner's syndrome—affected individual has only one X chromosome (XO)

   (c) Klinefelter's syndrome—affected person has extra X chromosome (XXY)

   (2) gene abnormalities

   (a) sickle-cell anemia

   (b) phenylketonuria (PKU)

   (c) sphingolipidoses (Tay-Sachs disease)

   (d) sex-linked diseases (hemophilia, others)

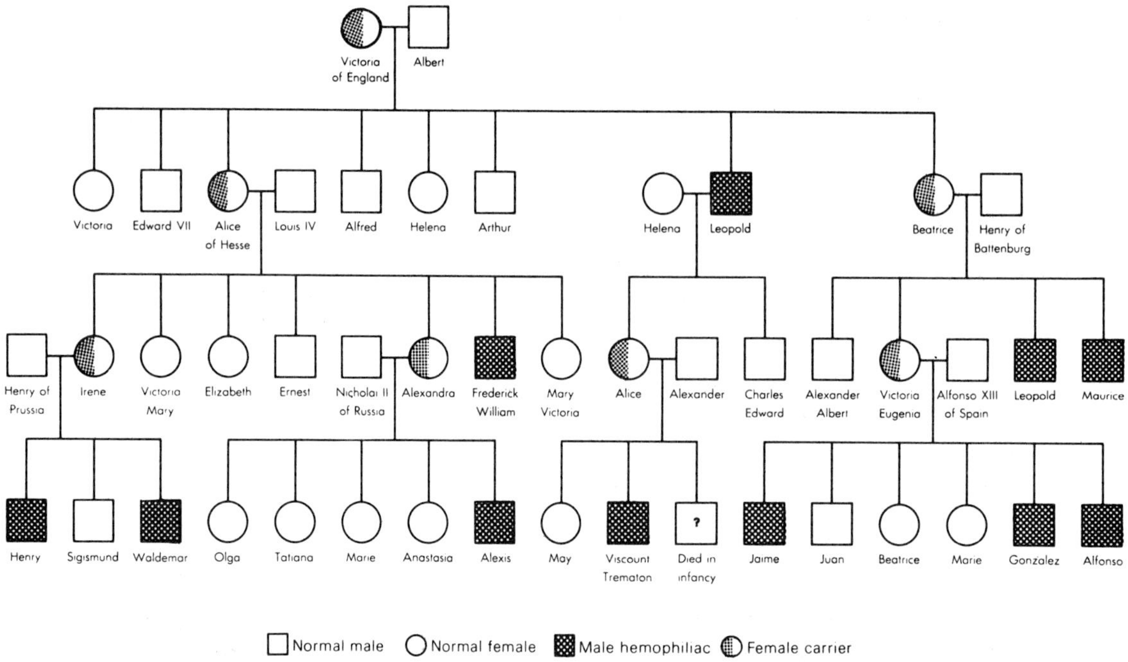

Figure 8. Pedigree chart of the relatives of Queen Victoria showing those members who suffered from hemophilia.

2. What part does diet play as a cause of disease?

Dietary diseases are caused by variations of food intake, either quality or quantity, that result in physiological deficiencies and changes.

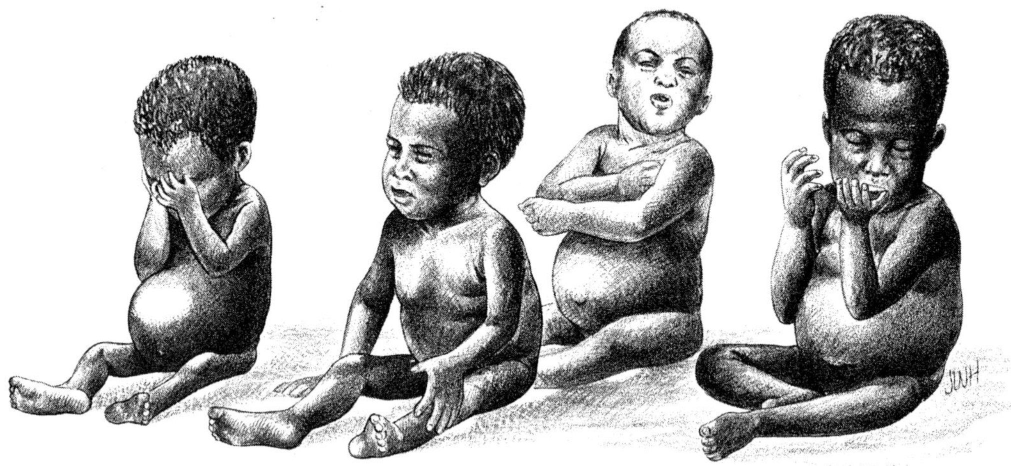

Figure 9. Children with the dietary disease kwashiorkor. (From Clark. 1973.)

EXERCISE 4. What are the differences between temperate and tropic zone diets?

(1) Study the data in Figure 10. What is the major difference between temperate zone and tropic zone diets?

(2) What effect does this difference have on protein requirements?

(3) What could be done to alleviate this condition in the tropical countries of the world?

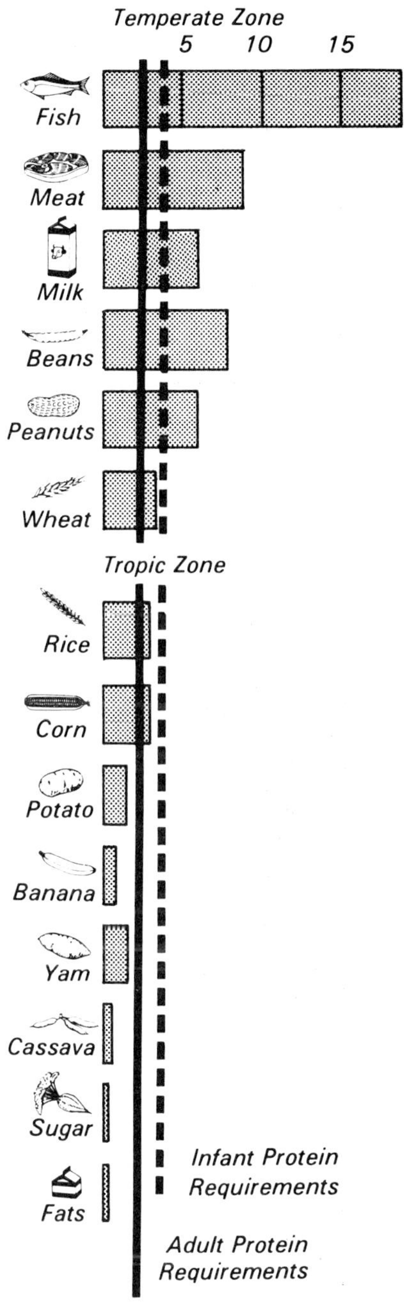

Figure 10. The relationship between infant and adult protein requirements and commonly eaten foods throughout the world. In this chart, typical foods eaten by people living in temperate zones are contrasted with typical foods in tropic zones. The numbers at the top indicate grams of protein per 100 calories of the various foods.

TABLE 3.

Common dietary diseases of man.

DIETARY DISEASE	CAUSE	PREVENTATIVE FOOD SOURCE
Kwashiorkor	protein deficiency	meat, fish
Scurvy	ascorbic acid (vitamin C) deficiency	citrus fruits, tomatoes, potatoes
Beriberi	thiamine (vitamin B_1) deficiency	whole grain cereals, liver, legumes
Ariboflavinosis	riboflavin (vitamin B_{12}) deficiency	whole grain cereals, green leafy vegetables, cheese, milk, eggs
Pellegra	niacin deficiency	liver, wheat germ, eggs, fish
Pernicious anemia	vitamin B_{12} deficiency	meat, liver, eggs, milk
Rickets	vitamin D deficiency	beef, butter, eggs, milk
Obesity	excess food	

3. What are the agents of infectious diseases, and how are the diseases transmitted?

TABLE 4.

Some common infectious diseases of man and the infecting agents.

DISEASE	INFECTING AGENT	DISEASE	INFECTING AGENT
Amoebic dysentery	Protozoan	Poliomyelitis	Virus
Athlete's foot	Fungus	Rheumatic fever	Bacterium
Bacterial endocarditis	Bacterium	Ringworm	Fungus
Bacterial pneumonia	Bacterium	Rocky Mountain spotted fever	Rickettsia
Bubonic plague	Bacterium		
Chagas' disease	Protozoan	Rubella (German measles)	Virus
Chancroid	Bacterium	Salmonellosis	Bacterium
Chickenpox	Virus	Scabies (the "itch")	Mite
Cholera	Bacterium	Scarlet fever	Bacterium
Common cold	Virus	Schistosomiasis	Liver fluke (a roundworm)
Conjunctivitis	Bacterium		
Diphtheria	Bacterium	Shigellosis	Bacterium
Enterobiasis	Pinworm	Sleeping sickness	Protozoan
Gonorrhea	Bacterium	Smallpox	Virus
Hepatitis	Virus	Syphilis	Bacterium
Histoplasmosis	Fungus	Tapeworm	Flatworm
Hookworm	Roundworm	Tetanus	Bacterium
Influenza	Virus	Tonsillitis (strep throat)	Bacterium
Leishmaniasis	Protozoan	Trichinosis	Roundworm
Lymphogranuloma venereum	Bacterium	Trichomoniasis	Protozoan
Malaria	Protozoan	Tuberculosis	Bacterium
Measles	Virus	Typhoid fever	Bacterium
Meningitis	Bacterium	Typhus	Bacterium
Mononucleosis	Virus	Undulant fever	Bacterium
Mumps	Virus	Viral pneumonia	Virus
Paratyphoid	Bacterium	Whooping cough	Bacterium
Pediculosis	Lice	Yellow fever	Virus

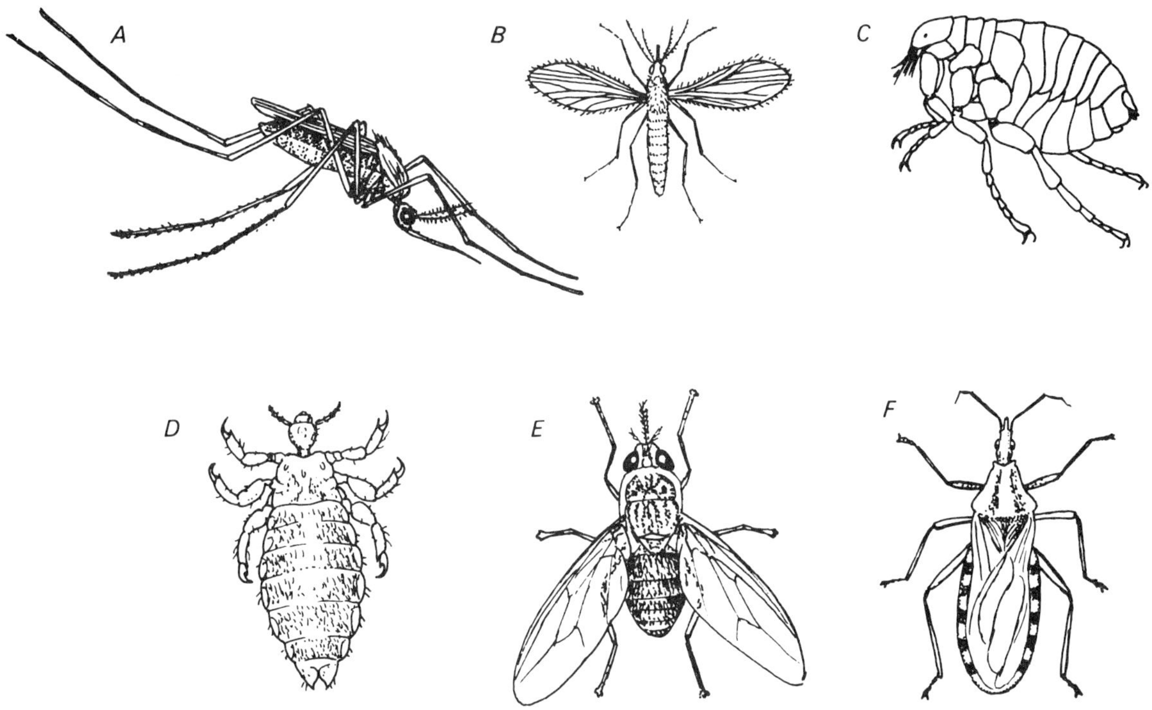

Figure 11. Six disease-carrying insects. *A*, The *Anopheles* mosquito (malaria); *B*, sandfly (leishmaniasis); *C*, rat flea (bubonic plague); *D*, body louse (typhus); *E*, tsetse fly (sleeping sickness); and *F*, kissing bug (Chagas' disease).

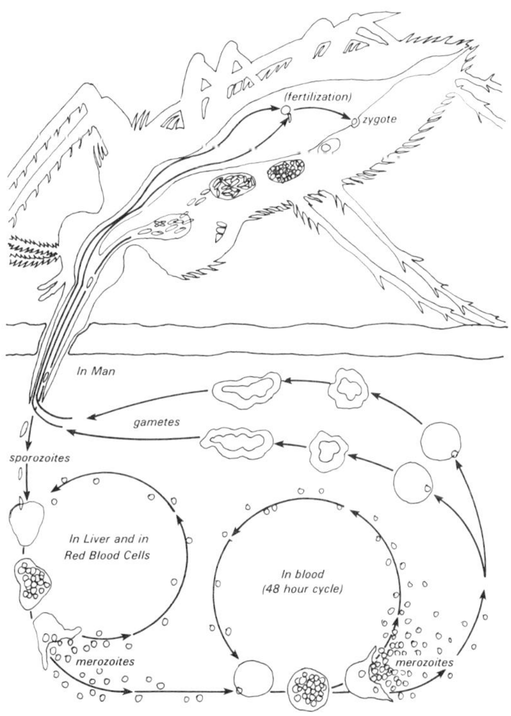

Figure 12. Life cycle of the malarial organism *Plasmodium vivax.* This sporozoan life cycle involves the *Anopheles* mosquito and man.

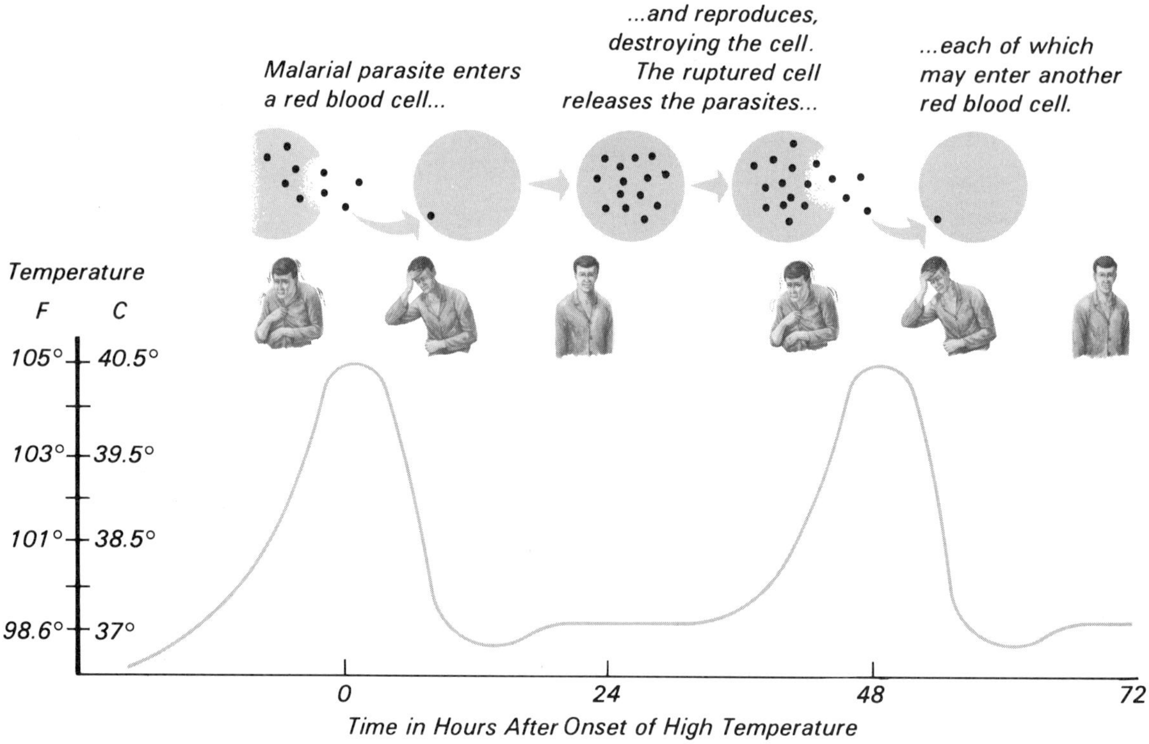

Figure 13. Temperature cycle in malaria, as related to life cycle of *Plasmodium.*

EXERCISE 5. How is malaria transmitted?

(1) What is the infectious agent of malaria?

(2) How is the disease transmitted to humans?

(3) Describe the occurrence of malaria in relation to the temperature cycle in humans and the life cycle of *Plasmodium.*

4. How do environmental factors cause disease?

EXERCISE 6. Anatomy of a disaster

Read the information in this exercise and study Figures 14 and 15. Then answer the questions at the end of the exercise.

Figure 14. View of Donora, Pennsylvania region. Industry and accompanying railroads line a curve of the Monongahela River, which is closely bordered by mountains on either side. (From Turk, et al. 1974.)

The circumstances surrounding the Donora disaster resembled those of the Meuse Valley [disaster in Belgium in 1930]. Large zinc smelters, plants producing wire, steel, and sulfuric acid, as well as many other factories, lined the river front solidly for three miles in the horsehsoe-shaped Monongahela Valley. Freight trains, operating day and night on both sides of the river and propelled by steam locomotives burning high-volatile soft coal, emitted clouds of smoke that were trapped by hills surrounding the river, some of which rise abruptly as high as 350 feet.

A high pressure condition has moved into western Pennsylvania, eastern Ohio, parts of Virginia, West Virginia, and Maryland. Rainfull had added fog and moisture to the valley's air. On October 26, the first day of the inversion, the weather was cold and wind velocity was near zero. A constant smog blanket produced a virtually airtight chamber, day and night, for more than five days, during which time the air was permeated by an odor of sulfur.

Twenty persons, seventeen in Donora and three in nearby Webster, died during the fateful days, and 5,910 persons, representing 42 percent of the people of that area, were stricken with irritation in eyes, nose, throat; pains and constriction in the chest; coughs, labored breathing, severe headaches, nausea and vomiting.

At a football game between Donora and Monongahela High Schools, players were obliged to drop out of the game because of chest pains, coughs, and shortness of breath. Patients improved temporarily when they stayed in their homes and closed their doors. According to health authorities, the older the Donorian, the more likely he was to become ill from the polluted air. Some of those exposed to it were permanently injured and subsequent deaths were attributed by the Public Health Service physicians to the air contamination. A veterinarian sent in by the Public Health Service reported that two dogs, seven chickens, three cats, two canaries, and two rabbits died from the smoke. According to a follow-up study ten years later, Donorians who became acutely ill at the time of the smog episode showed a higher mortality and illness prevalence than others who were residing in the town at that time. The highest illness rates following the disaster were among smokers who had become ill during the episode; the lowest rates were in nonsmokers who were not ill during the inversion.

From Waldbott, George L., Health Effects of Environmental Pollutants, *St. Louis, 1973, The C. V. Mosby Company.*

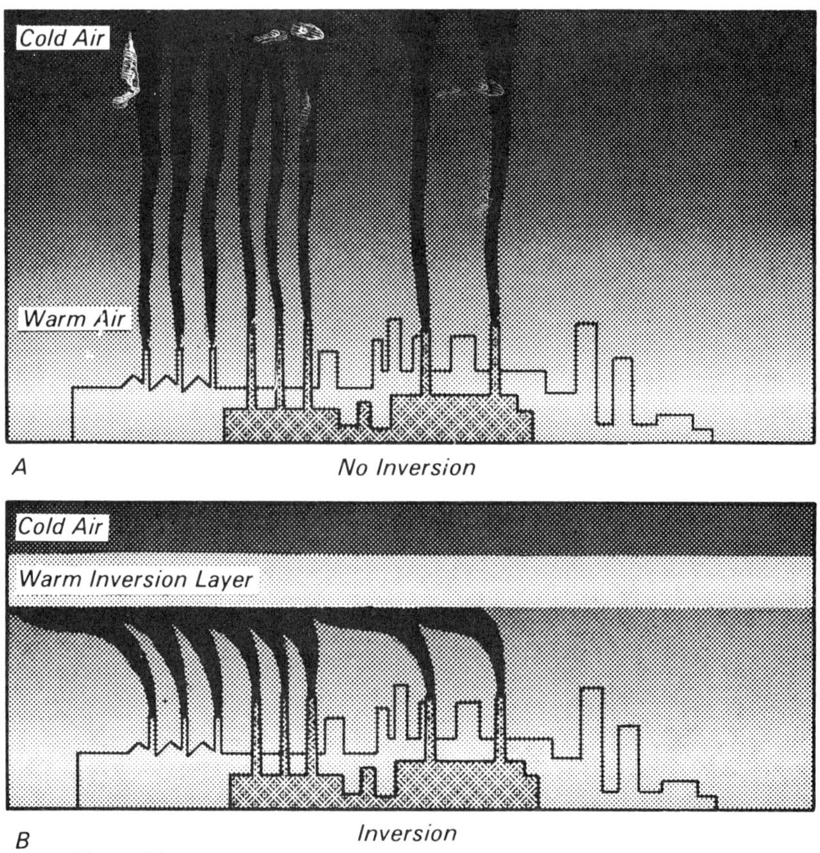

Figure 15. Weather effects on smog formation. *A,* Normally warm air from a city rises, carrying pollutants with it. *B,* When temperature inversions occur, a layer of warm air traps pollutants beneath. (From Clark. 1973.)

(1) List below the two main factors that contributed to this disaster.

(2) List below any specific conditions that were under human control.

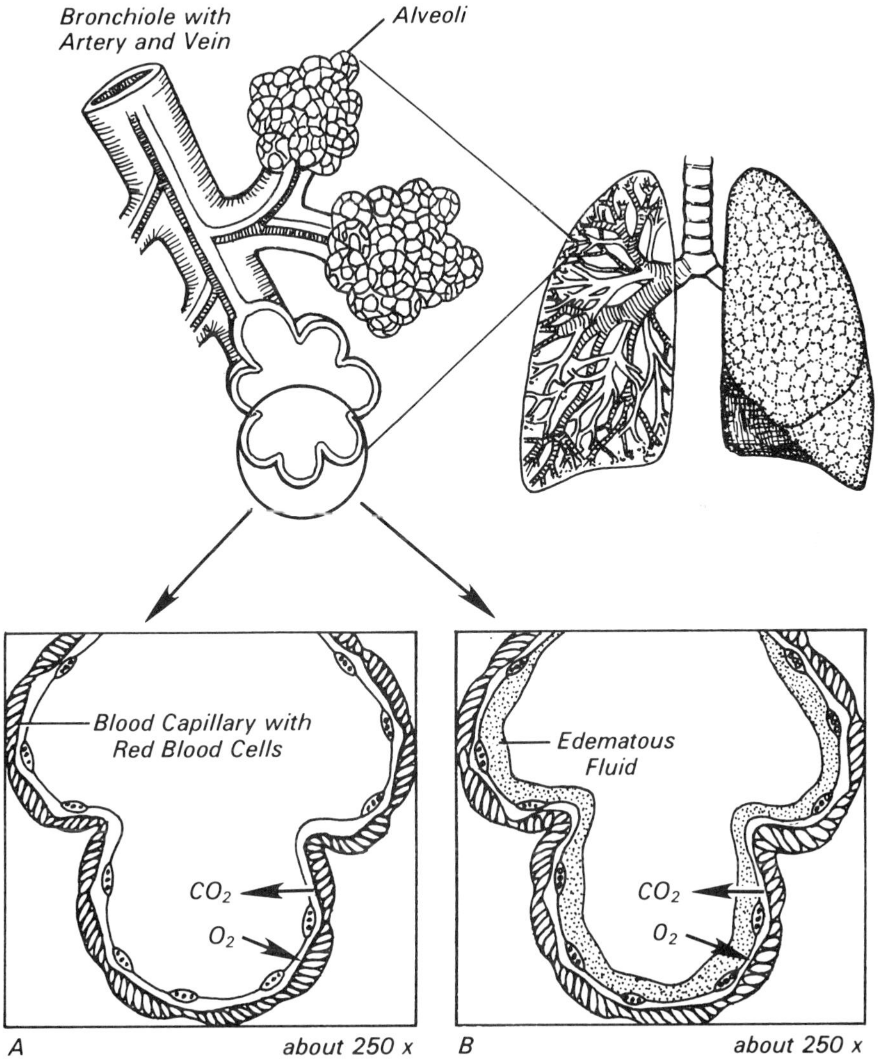

Figure 16. Diagram of human lung, showing circulation of air and blood to alveoli. *A,* Normal lung—gases exchange readily between air spaces and blood, passing across extremely thin cells lining alveoli. *B,* Edematous lung—alveoli are partially filled with fluid, whicn greatly increases the distance across which gases must diffuse, slowing down the rate of exchange. (From Clark. 1973.)

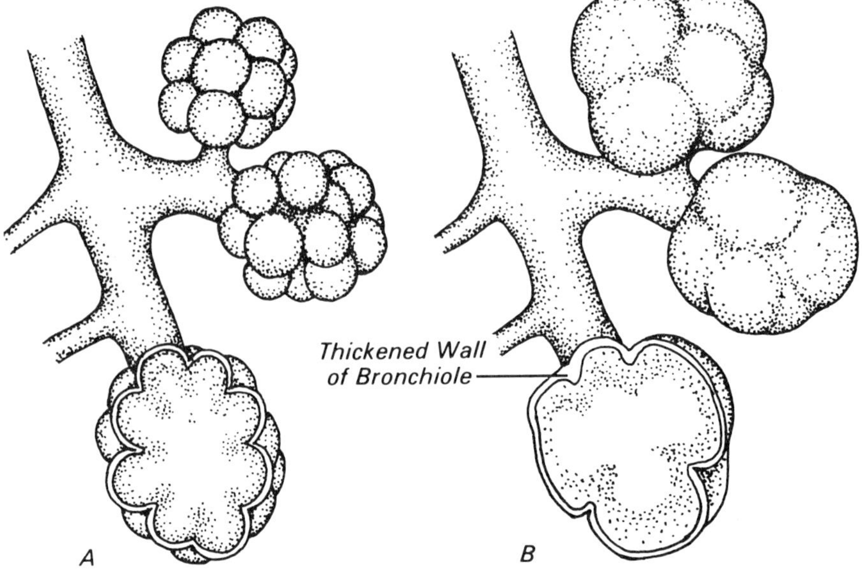

Figure 17. Diagram of human lung showing breakdown of alveolar walls during development of emphysema. *A*, Normal lung. *B*, Lung with emphysema—alveolar walls have disappeared, decreasing the surface available for gas exchange. The walls of the bronchioles have become thick and less elastic and are more resistant to the flow of air. (From Clark. 1973.)

5. stress—any stimulus that interferes with a person's normal physical, mental, or emotional well-being.

6. degenerative processes

    a. degenerative arthritis

    b. emphysema (certain forms)

    c. cataracts

    d. cancer (certain kinds)

    e. hypertension

    f. arteriosclerosis

    g. cerebrovascular disease (stroke)

    h. heart disease

AIR POLLUTION DISASTERS

Meuse Valley, France, 1930
63 dead
Donora, Pennsylvania, 1948
20 dead
5,910 ill
London, 1952
4,000–5,000 dead
New York, 1953
?? dead

LOS ANGELES COUNTY

AIR POLLUTION CONTROL DISTRICT

434 South San Pedro Street, Los Angeles, California 90013/629-4711

LOUIS J. FULLER
Air Pollution Control Officer

ROBERT L. CHASS
Chief Deputy

To All Schools in the Los Angeles Basin

Gentlemen:

On March 3, 1969, the Los Angeles County Medical Association unanimously adopted the following resolution:

> "Because SMOG is an increasing health hazard which may seriously affect the lungs of young people, the Committee on Environmental Health of LACMA strongly recommends that when the forecast concentration of OZONE (oxidants) in the atmosphere reaches 0.35 part per million, Los Angeles County students through high school, in any identified air monitoring zone, should be excused from strenuous indoor and outdoor activities. The OZONE (oxidants) information will be transmitted by the Air Pollution Control Officer to the Board of Education in the specific zones affected by the air pollution."

At its meeting on April 1, 1969, the Los Angeles County Board of Supervisors adopted an order directing the Air Pollution Control Officer to implement the above resolution.

The news media are the most practical method of transmitting forecast information on ozone levels. Most of the County's radio, television, and press will carry this information as a "School Smog Warning" from the time the forecast is released (afternoon before the school day), and, in addition, the following radio stations, among others, will carry the warning on their news broadcasts between the hours of 7 a.m. and 9 a.m.: KNX, KFWB, KFI, KLAC, KGFJ, KPOL, KABC, KGIL, and KHJ.

School Smog Warnings will be carried by the news media only when the Air Pollution Control District forecasts ozone levels to reach or exceed the level contained in the resolution.

Any questions concerning health effects should be directed to the Los Angeles County Medical Association. Information regarding forecasts may be obtained from the District's Public Information and Education Division, 629-4711, extension 66101.

Very truly yours,

Louis J. Fuller
Air Pollution Control Officer

LJF:sw

Attachments

Figure 18. A letter from the Air Pollution Control Officer to all Los Angeles schools, April 1969. (From Clark. 1973.)

V. WHAT METHODS ARE AVAILABLE FOR THE CONTROL AND PREVENTION OF DISEASE?

A. How Does the Physician Diagnose a Disease?

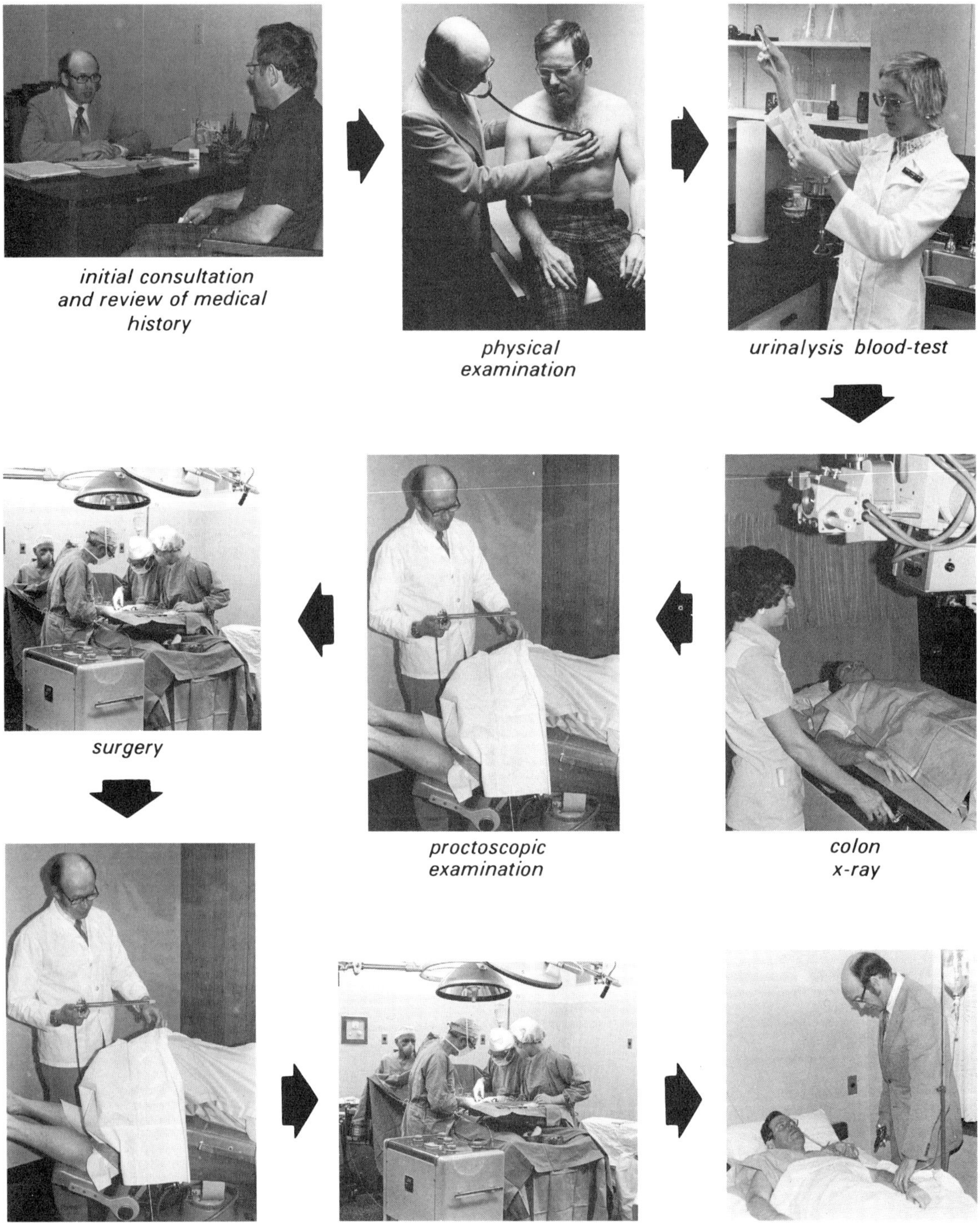

Figure 19. Steps in the diagnosis and treatment of a disease.

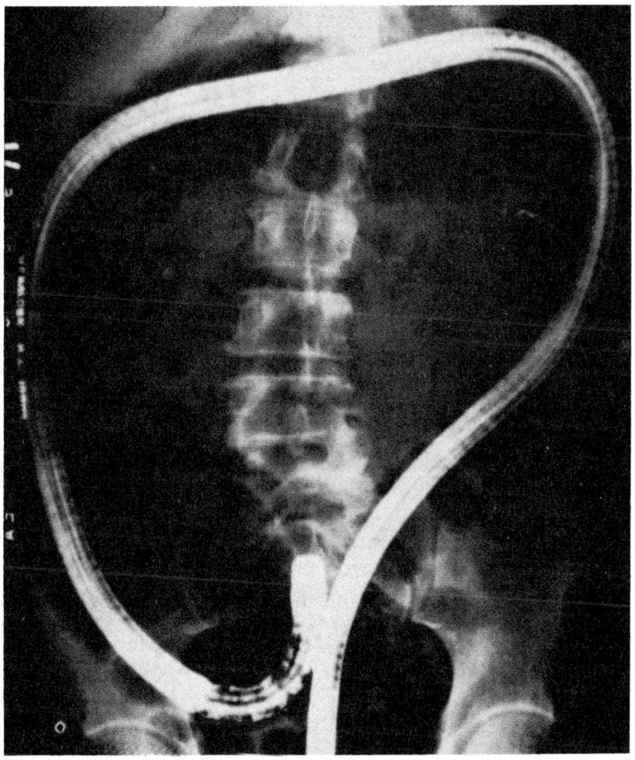

Figure 20. Use of the colonoscope. The colonoscope is inserted into the colon (large intestine). It allows the doctor to see a full six feet up the colon, and to remove polyps from the colon before they become cancerous (From Arehart-Treichel. 1975. Photograph courtesy Dr. William I. Wolff and Dr. Hiromi Shinya, Beth Israel Medical Center, New York.)

B. Technological Tools of the Physician

1. chemotherapy
   a. infectious diseases
   b. degenerative diseases
   c. mental disorders
2. surgery
   a. removal
   b. repair
   c. replacement

C. Frontier Procedures and Technologies

1. closed-circuit television
2. computers
3. human care space satellites
4. telemedicine networks
5. acupuncture
6. molecular biology and genetics

HAS THE BATTLE BEEN WON?

APPENDIX I
Answers to Exercises

EXERCISE 3.

Possible answers:

TABLE 5.

The leading causes of death in the United States in 1972.*

CAUSE OF DEATH	DISEASE-CAUSING FACTOR(S)
1. Disease of the heart	1, 2, 4, 5, 6
2. Cancer	1, 2, 4, 6
3. Cerebrovascular disease (stroke)	1, 2, 4, 5, 6
4. Accidents	– –
5. Influenza and pneumonia	3, 4
6. Diabetes mellitus	1, 2, 6
7. Symptoms and ill-defined symptoms	– –
8. Diseases of infancy	1, 2, 3, 4
9. Arteriosclerosis**	1, 2, 5, 6
10. Cirrhosis of liver	1, 2
11. Suicide	– –
12. Emphysema	4, 6
13. Homicide	– –
14. Congenital anomalies	1
15. Nephritis and nephrosis†	3, 6
16. Ulcer of stomach and duodenum	2, 4, 5
17. Hypertension	4, 5, 6
18. Infections of kidney	3
19. Hernia and intestinal obstructions	2, 5, 6
20. Bronchitis	3, 4
21. Tuberculosis	3, 4
22. Other infective and parasitic diseases	3

*Data from the Department of Health, Education, and Welfare, Public Health Service, Annual Report, *Vital Statistics of the United States.*

**Medical term for hardening of the arteries.

†Nephritis, also called Bright's disease, is the medical term for inflammation of the kidneys; nephrosis is a term for noninflammatory degeneration of the kidneys.

Many diseases have indirect causes that are difficult to identify. Some of the causes listed in this table are speculative and should not be considered absolute.

EXERCISE 5.

(3) The *Plasmodium* sporozoites live in the salivary glands of the mosquito (genus *Anopheles*). When the mosquito bites a human, it injects saliva, which contains a chemical that prevents coagulation (clotting) of the human blood and permits the insect to suck up a supply of blood on which to feed. The sporozoites enter the human bloodstream with the mosquito saliva and move into the red blood cells. In the blood cells, the sporozoites undergo a period of growth and development, followed by a series of mitotic divisions that produce large numbers of merozoites. When the blood-cell wall disintegrates, the merozoites move passively through the blood and enter other blood cells, where the process of asexual reproduction is repeated to produce still more merozoites. The liberation of merozoites from blood cells is synchronized throughout the bloodstream, occurring approximately every 48 hours. Toxic substances released along with the merozoites cause the recurrent fevers that typify malarial attacks.

ADDITIONAL INQUIRIES

1. Compare the effectiveness of several antibiotics in the control and inhibition of growth in different bacteria cultures.

2. Arrange to visit your doctor or a nearby hospital to discuss and examine the array of medical equipment used in diagnosis and treatment of diseases in various categories, i.e., infectious, congenital, stress, dietary, degenerative, and environmental.

3. Make a library inquiry for actuarial tables of the kind used by life insurance companies. Look up data by height and weight. What have these data to do with life span? How do they correlate with particular diseases studied in this minicourse?

4. Additional readings:

Arehart-Treichel, Joan. 1975. Is the great cancer cure already with us? Science News. *107*:26–27.

Thorwald, Jurgen. 1957. The century of the surgeon. Pantheon Books, New York.

GLOSSARY

ALVEOLI: The tiny air sacs in the lungs through which exchange of gases occurs.

CHEMOTHERAPY: The use of chemicals in the treatment or control of disease.

DISEASE: Any abnormal functioning of any structure, organ, or system of the body.

EPIDEMIOLOGY: The science of medicine dealing with the incidence, distribution, and control of disease in a population.

GAMETE: A mature reproductive cell; a sperm or egg.

GAMETOCYTE: A cell that gives rise to gametes.

HEALTH: A state of complete physical, mental, and social well-being and not merely the absence of disease and infirmity.

INFECTION: A condition produced by the establishment of a disease-producing substance or agent in a host organism.

MEROZOITES: A form of spores produced by segmentation of the schizonts.

NUTRITION: The relationship of the health and well-being of an organism to the food it consumes.

OOCYST: An encysted spore after fertilization.

PROCTOSCOPIC EXAMINATION: The use of a dilating instrument to examine the anus and rectum.

SCHIZONTS: Cells formed by the growth of sporozoites in the host.

SPOROZOITES: Small, usually motile, infective stages of a protozoan introduced into a host.

SYMPTOM: A change in the body or its functions, which is indicative of a disease.

TECHNOLOGY: The application of scientific knowledge to advance the state of any applied science such as medicine.

TEMPERATURE INVERSION: A condition in which a layer of warm air above cooler air prevents the vertical movement and mixing of air from the earth's surface.

ZYGOTE: A fertilized egg.

REFERENCES

Arehart-Treichel, J. 1975. Is the great cancer cure already with us? Science News. *107*:26–27.

Clark, M. E. 1973. Contemporary biology: concepts and implications. W. B. Saunders Company, Philadelphia. pp. 229–243.

Communications Research Machines Books. 1972. Life and health. Communications Research Machines, Inc., Del Mar, California.

Diehl, H. S., and W. Dalrymple. 1968. Healthful living. Eighth Edition. McGraw-Hill Book Company, New York.

Dubos, R., and M. Pines (eds.). 1965. Health and disease. Time-Life Books, Inc., New York.

Kogan, B. A. 1974. Health: man in a changing environment. Second Edition. Harcourt Brace Jovanovich, Inc., New York.

Smith, D. W., and A. A. Wilson. 1973. The child with Down's syndrome. W. B. Saunders Company, Philadelphia.

Taylor, W. T., and R. J. Weber. 1968. General biology. Second Edition. D. Van Nostrand Company, Princeton, New Jersey.

Turk, A., et al. 1974. Environmental science. W. B. Saunders Company, Philadelphia.

Waldbott, G. L. 1973. Health effects of environmental pollutants. C. V. Mosby Company, St. Louis.

study guide

the biological basis of race

minicourse

WHAT DO YOU
ASSOCIATE WITH RACE?

PREREQUISITES: An understanding of natural selection and Mendelian genetics.

STATEMENTS ABOUT RACE—TRUE OR FALSE?

T	F		
___	___	1.	The history of the human species indicates a migration from hot equatorial regions to more temperate and cooler regions. As a consequence, lighter skin may have evolved through natural selection.
___	___	2.	The heavily pigmented skin of the Black tends to prevent the penetration of harmful sun rays.
___	___	3.	Significant proof exists that certain races are intellectually superior to others.
___	___	4.	The gene pools of human races differ by less than one percent.
___	___	5.	The various human races existing at the present time have been in existence since shortly after humans evolved.
___	___	6.	All people living on Earth today can be classified into three races.
___	___	7.	There is a relationship between some characteristics of groups and the race(s) to which they belong.
___	___	8.	Every individual can be classified as belonging to a given race by means of certain characteristics.
___	___	9.	During the last several centuries a significant decrease in the number of races has occurred as a result of migration.
___	___	10.	American Blacks compose a race.
___	___	11.	Races result only from geographical isolation.
___	___	12.	Jews compose a race.
___	___	13.	If the only characteristic you could observe was skin color, you could use this as a reliable characteristic to identify a person as a member of a particular race.
___	___	14.	The child of a white and black couple will always have skin color midway between its parents.
___	___	15.	Skin color is entirely a genetic characteristic.

I. WHAT IS RACE?

A. How Does Race Fit Into the Biological Classification Scheme?

Figure 1. The people of the world. How many "races" are there? (Photograph courtesy Monkmeyer Press Photo Service.)

TABLE 1.

Categories of classification of living organisms.

CATEGORY	EXAMPLES	
	Humans	Housefly
Kingdom	Animalia	Animalia
Phylum	Chordata	Arthropoda
Class	Mammalia	Insecta
Order	Primates	Diptera
Family	Hominidae	Muscidae
Genus	*Homo*	*Musca*
Species	*Homo sapiens*	*Musca domestica*
Subspecies, Races, Varieties		

A race is a breeding population of individuals, identifiable by the frequency with which a number of inherited traits appear in that population.

A species is a group of actually or potentially interbreeding natural populations that is reproductively isolated from other such groups.

B. How Is the Concept of Race Useful in Biology?

1. Domesticated races are designed by man for his practical use.

   a. cattle

Holstein *Santa Gertrudis* *Hereford*

Figure 2. The three categories of cattle pictured above have been bred for specific purposes. The Holstein (*left*) is an excellent milk producer; the Santa Gertrudis (*center*) is an efficient meat producer on dry open range; and the Hereford (*right*) is an excellent meat producer on lush pastures.

   b. dogs of the world

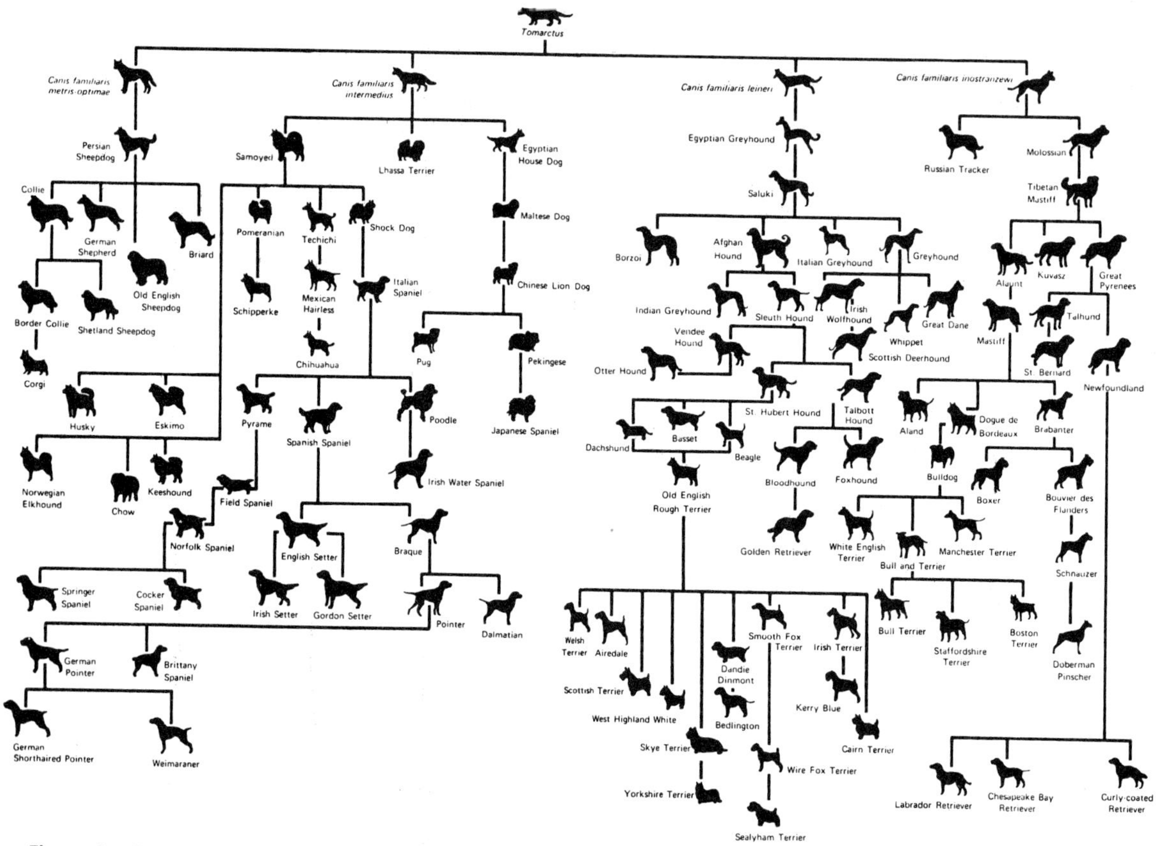

Figure 3. Dog breeds of the world. (Reprinted with permission of Macmillan Publishing Company, Inc., from *Race and Races* by R. A. Goldsby. Copyright © 1971 by Richard A. Goldsby.)

2. What is the relationship of race to geography?

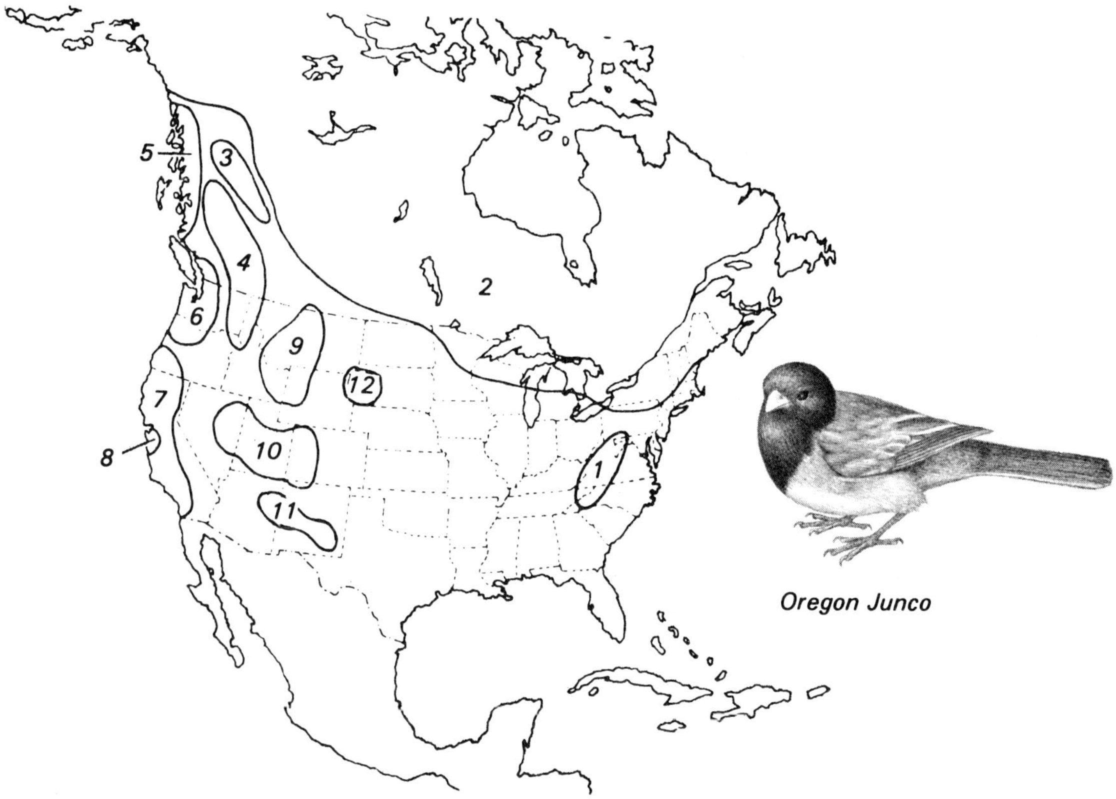

Figure 4. Map showing the breeding areas of twelve subspecies of the North American junco (*Junco hyemalis*), a typical polymorphic species. (1) *Junco hyemalis carolinensis;* (2) *J. hyemalis hyemalis;* (3) *J. h. cismontanus;* (4) *J. h. montanus;* (5) *J. h. oregonis;* (6) *J. h. shufeldti;* (7) *J. h. thurberi;* (8) *J. h. pinosus;* (9) *J. h. mearnsi;* (10) *J. h. caniceps;* (11) *J. h. dorsalis;* and (12) *J. h. aikeni.* (From *The Biology of Race* by J. C. King, Copyright 1971 by Harcourt Brace Jovanovich, Inc., and reproduced with their permission.)

Chief Thundercloud An American Indian — *A Young Chinese Woman* — *A Greenland Eskimo*

Figure 5. Representatives of three races of similar oriental ancestry. *Left to right:* Chief Thundercloud, an American Indian; a young Chinese woman; and a Greenland Eskimo. (Photographs courtesy the Peabody Museum, Harvard University.)

II. HOW DO RACES DEVELOP BIOLOGICALLY?

A. Species, Races, and Their Gene Pools

1. How can genetic drift account for variations in gene pools?

Gene Pool: *The sum of all the genes in an interbreeding population.*

EXAMPLE: A, B, AB, and O Blood Types

Alleles: A, B, i

Possible Genotypes: Type A Blood (AA, Ai); Type B Blood (BB, Bi): Type AB Blood (AB); and Type O Blood (ii).

ABi Alleles in Gene Pool: Sum of the A, B, i alleles in the population, expressed in terms of the frequency or percentage of each allele in the total.

If there are 100 blood type alleles in the pool (50 individuals in the population), and 20 are the A allele, then the frequency of the A allele is 0.2 (20 percent).

EXERCISE 1. Possible evolution of a gene change

BAIB is a chemical compound excreted in the urine of some people. It has no known function in man. Read the hypothetical case study below and then answer the questions that follow it.

A HYPOTHETICAL CASE STUDY

Consider a time 20,000 or 30,000 years ago when human populations were sparsely distributed collections of small tribes. Suppose a tribe of 45 members is made up of 20 adults and 25 children. Suppose further that 15 of the children carry the gene for high BAIB excretion and 10 do not. In this situation, 60 percent of those children who will produce the next generation carry this gene and 40 percent do not. Suppose some tigers surprised a group of these children at play and killed, just by chance selection, 5 children who did not carry the BAIB excretion gene. The future generations of this tribe will now be drawn from a pool whose frequency is 75 percent excretor gene, 25 percent nonexcretor gene. The gene frequency for the BAIB excretion has suddenly drifted upward solely because some nonexcretors happened to be in the wrong place at the wrong time. Any shifts in the frequency of adaptive or deleterious genes could be brought back into environmental harmony by the pressures of natural selection. However, because we have declared the BAIB excretion genes neutral, they will not be affected by natural selection.

(1) List three circumstances that might be factors involved in changing gene frequencies in this primitive population.

(2) Could these types of changes occur in present world populations?

2. What effects does the environment have on gene frequencies?

Many changes in the makeup of a gene pool are a result of the adaptability of organisms that possess certain advantageous genetic characteristics.

EXERCISE 2. Natural selection and changes in gene pools

Suppose another primitive population is a large one, and that 60 percent of its members have the gene for high BAIB excretion while 40 percent do not. Assume also that BAIB excretors require a higher water intake than nonexcretors. Describe the effect on the gene pool of the population if climatic changes tend increasingly to dryness and drought.

B. What Is the Effect of Isolation on Race Formation?

EXERCISE 3. Gene pool separation

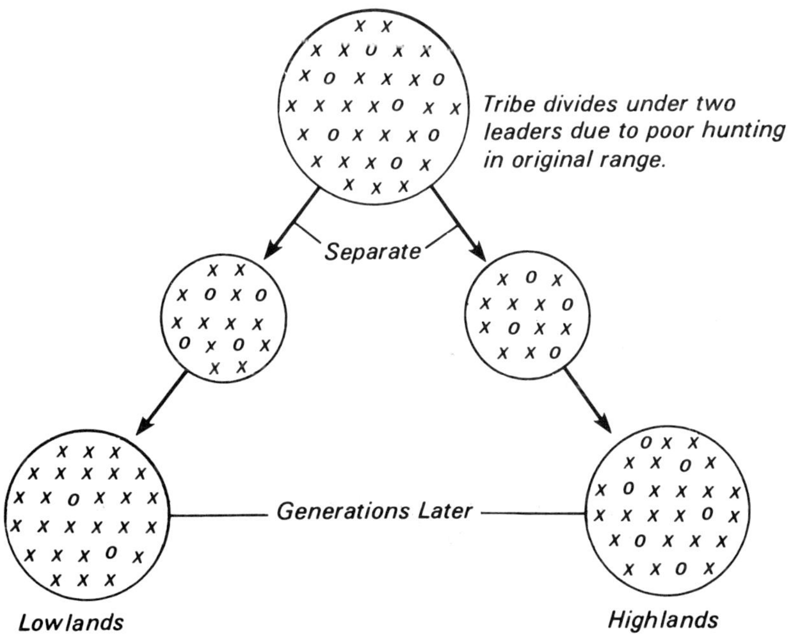

Figure 6. Separation of a hypothetical gene pool.

The members of a tribe (population) of hunters with either X or O characteristics divide under two leaders because of poor hunting in their original range. One group settles in the lowlands, the other in the highlands.

In time, what might cause the two populations to differ?

C. How Does Natural Selection Affect Race Formation?

EXERCISE 4. Variation in limb/weight ratios of selected human populations

TABLE 2.

Limb length/body weight ratios (cm/kg) for adult males. Numbers in the left-hand column refer to the number of individuals in each sample. The far right column lists approximate latitudes.*

NUMBER OF INDIVIDUALS AND POPULATION DESCRIPTION	AVG. LIMBS/ WEIGHT RATIO	APPROX. LATITUDE	NUMBER OF INDIVIDUALS AND POPULATION DESCRIPTION	AVG. LIMBS/ WEIGHT RATIO	APPROX. LATITUDE
73 Parisian workers	4.88	50° N	119 Nhungues (Mozambique)	5.66	15° S
47 Finns	4.89	60° N	123 Darod (Somalia)	5.74	5° N
80 French soldiers	4.91	50° N	119 Rahanoween (Somalia)	5.83	5° N
50 French students	4.94	50° N	51 Gobawoen (Somalia)	5.90	5° N
300 British soldiers**	5.00	55° N	47 Dir (Somalia)	6.01	5° N
113 French soldiers	5.02	50° N	26 Antumba (Mozambique)	6.06	15° S
504 Ukrainians**	5.08	50° N	87 Hawyah (Somalia)	6.21	5° N
120 Sicilian soldiers	5.09	40° N	18 Korana (South Africa)	6.21	25° S
100 Tonkinese	5.37	20° N	35 Indians (Madras)	6.61	10° N
82 Otomis (Mexico)	5.51	20° N	95 Aka Pygmies (Congo)	6.98	5° S
31 Arabs (Yemen)	5.63	20° N	115 Basua Pygmies (Congo)	7.03	5° S
18 Asheraf (Somalia)	5.64	5° S			

*From Schreider, 1957 and unpublished data. Data concerning Finns and Madras Indians communicated by M. Pelosse (Paris) and Dr. Henrotte (Liege), respectively.

**Calculated from published averages. All other figures calculated from individual measurements.

Evaluate the data in Table 2. Study the characteristic of limb length/body weight ratio in relation to distance north or south of the equator.

(1) What is the general relationship between the limb length/ body weight ratio and where people live, relative to the equator?

(2) What conclusion might be drawn about the relationship between warmer or colder climates and length of limbs in relation to body weight?

(3) How do these data suggest that natural selection has affected these populations?

D. How Can Cultural Pressures Affect Race Formation?

EXERCISE 5. Factors influencing human gene pools

Read the four numbered examples and answer the question that follows them.

(1) albinism in Hopi Indian tribes

It has been hypothesized that albinism increased among the Hopi Indians of the southwest United States because of a social custom. Albino males could not work in the fields and travel with the men of the tribe. They stayed at home with the women and grew to occupy a special status, leading to relative freedom to mate with the tribe's women. This increased their numbers in the tribe.

(2) changes in selective pressure as a result of forced migration

In the United States, sickle-cell anemia is declining in the Black population because of a change in selective pressure. In the tropics in Africa, where sickle-cell anemia is prominent, the carriers of the gene have a higher resistance to malaria. In the United States, no such positive pressure is present to offset the negative effects of this genetic disease. Therefore, only a negative pressure exists, and a corresponding decrease in sickle-cell anemia has occurred.

(3) isolation by artificial barriers

Cultural barriers have for years isolated groups from one another—Whites from Indians, Irish from Jews, Chicanos from Blacks, Orientals from Whites, and so forth. These separations have tended to continue isolating the groups and to slow down the process of forming a separate set of racial characteristics within the United States.

(4) legal examples that produce isolation

The following legal definitions have been used to define races:

This description of racial identity as a Jew was contained in the anti-Semitic laws of Nazi Germany. It appeared as the first regulation of the Reich Citizenship Law.

A Jew is anyone who descended from at least three grandparents who were racially fully Jews. A Jew is also one who descended from two full Jewish parents if (a) he belonged to the Jewish religious community at the time this law was issued or he joined the community later; (b) he was married to a Jewish person at the time this law was issued or married one subsequently; (c) he is the offspring from marriage with a Jew contracted after the law for the protection of German Blood and German Honor became effective; (d) he is the offspring of an extramarital relationship with a Jew and was born out of wedlock.

Notice the absence of actual genetic information. In the Law for the Protection of German Blood and German Honor, you can see some of the consequences that flow from a political rather than a biological definition of "race." The second of the Nazi race laws decreed the following:

Permeated by the knowledge that the purity of the German blood is the basis for the permanence of the German people and animated by the inflexible determination to safeguard the German nation for all time, the Reichstag has unanimously decreed the following law:

1. *Marriages between Jews and citizens of German or similar blood are forbidden.*
2. *Extramarital intercourse between Jews and citizens of German and similar blood is forbidden.*
3. *Jews may not employ female citizens of German and similar blood under forty-five years of age in their households.*

Germany enacted the laws that contained these discriminations against a nonexistent race during the 1930's, in an orgy of hysteria. Wild schemes for the protection of "German blood" from Jewish "pollution" were common. One of the most bizarre of these involved using a pendulum as a sort of divining rod to detect individuals with "Jewish" blood.

Laws aimed at protecting imagined "racial purity" have not been the exclusive property of Nazism. Until the United States Supreme Court ruled them unconstitutional in 1966, miscegenation (a mixture of races) laws designed to protect the "purity" of the White race were part of the legal codes of many states in this country. Here are some examples of statues containing legal definitions of identity as a Negro:

1. A miscegenation statute of Missouri decreed it was a felony for Whites to marry Negroes. *"Negroes are defined as persons having one-eighth part or more of Negro blood."*
2. An Arkansas statute stated that it was a misdemeanor for any White person to marry *"any person who has in her or his veins any Negro blood whatsoever."*
3. A Texas miscegenation law stated that it was a felony for Whites to marry *"Africans or descendants of Africans."*

Explain how these four different examples all illustrate a form of isolation, but with different results. Write your answers in the margin beside each example.

III. WHAT HUMAN POPULATIONS ARE CONSIDERED RACES AND HOW SIGNIFICANT ARE THE DIFFERENCES BETWEEN THEM?

A. What Are the Human Races?

1. races of the world

1. Northwest European
2. Northeast European
3. Alpine
4. Mediterranean
5. Hindu
6. Turkic
7. Tibetan
8. North Chinese
9. Classic Mongoloid
10. Eskimo
11. Southeast Asiatic
12. Ainu
13. Lapp
14. North American Indian
15. Central American Indian
16. South American Indian
17. Fuegian
18. East African
19. Sudanese
20. Forest Negro
21. Bantu
22. Bushman and Hottentot
23. African Pygmy
24. Dravidian
25. Negrito
26. Melanesian-Papuan
27. Murrayian
28. Carpentarian
29. Micronesian
30. Polynesian
31. Neo-Hawaiian
32. Ladino
33. North American Black
34. South African Black

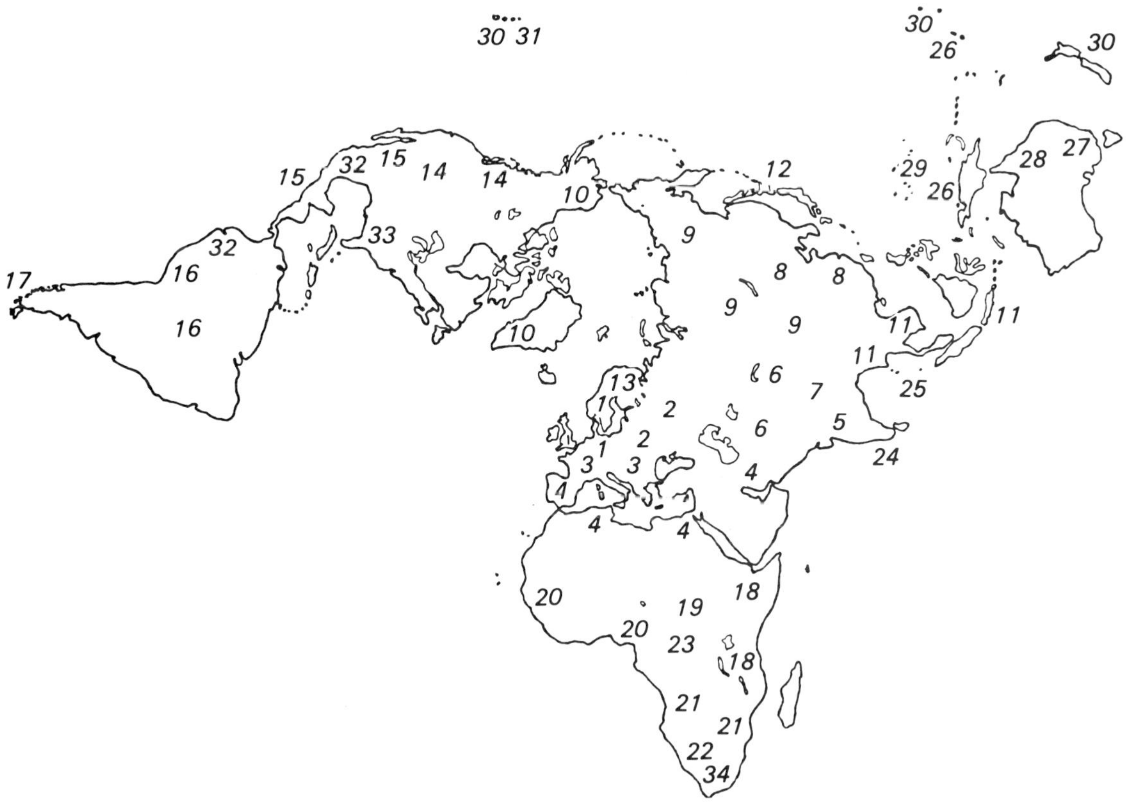

Figure 7. The geographical occurrence of a majority of thirty-four races of man. (From Dobzhansky. 1962.)

2. the large groups
   a. Caucasian
   b. Negroid
   c. Mongoloid
   d. Indian subcontinents

B. What Characteristics Can Be Used to Identify Races?

1. some examples

EXERCISE 6. What are some selected racial characteristics?

General Directions:

(1) With a cotton swab, remove a small quantity of wax from your ear (be careful not to push too hard and damage the ear). Is it dry and crumbly, or moist and adhesive? Compare your results with the information in Table 3. Record your observations in Table 7.

TABLE 3.

Percentage of population bearing the *dry* ear wax gene.*

POPULATION EXAMINED	PERCENTAGE BEARING GENE
Northern Chinese	98
Southern Chinese	86
Japanese	92
Melanesians	53
Micronesians	61
Germans	18
American Whites	16
American Blacks	7

*Data from Matsunaga. 1962.

(2) Taste a short strip of PTC paper. Tasters can detect a sharp bitter taste immediately. Compare your results with Table 4 and record them in Table 7.

TABLE 4.

Percentage of PTC tasters in different populations.*

POPULATION	PERCENTAGE OF TASTERS
Europeans	59–83
Negroes	91–97
East Asians (Chinese, Japanese, and related populations)	83–100
Australian Aborigines	49–67
Micronesians	72–82

*Data from Coon. 1965.

(3) Match your fingerprint pattern with one of the three in Figure 8. Compare your findings with the information in Table 5. Record your observations in Table 7.

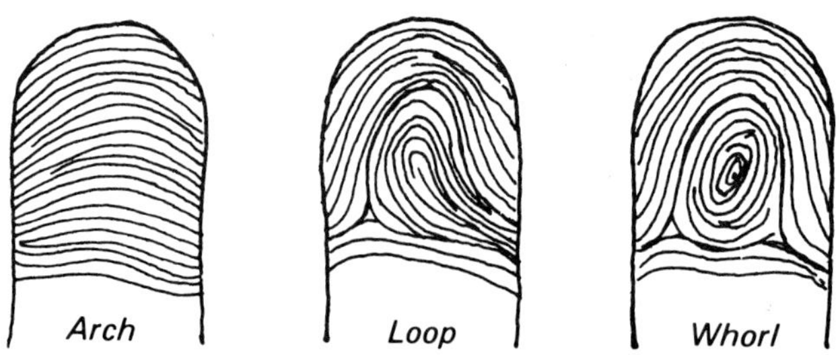

Figure 8. Three basic types of fingerprints.

TABLE 5.

Percentage range of types of fingerprints among populations.*

POPULATION	ARCHES	LOOPS	WHORLS
European	0–9	63–76	20–42
Negroes	3–12	53–73	20–40
Bushmen	13–16	66–68	15–21
East Asians (Chinese, Japanese, and related populations)	1–5	43–56	44–54
American Indians	2–8	46–61	35–57
Australian Aborigines	0–1	28–46	52–73
Micronesians	2	49	49–50

*Data from Coon. 1965.

(4) Most people know their blood type, but if you do not, your instructor can help you with a quick typing. Compare your type for the A, B, O system and the Rh factor with the data in Table 6. Record your results in Table 7.

TABLE 6.

Blood-group variation (percentages) among populations.*

POPULATION	A	AB	B	O	Rh NEGATIVE	DUFFY FACTOR
Caucasians	5–40	1–37	4–18	45–75	25–46	37–82
Negroes	8–30	1–8	10–20	52–70	4–29	0–6
East Asians	0–45	0–5	16–25	39–68	0–5	90–100
American Indians	0–20	about 0	0–4	68–100	about 0	22–99

*Data from Coon. 1965.

TABLE 7.

Student data table.

CHARACTERISTIC	MY RESULTS	POSSIBLE POPULATIONS
1. Ear Wax		
2. PTC		
3. Fingerprint		
4. Blood group		

Questions:

(1) How well does your set of characteristics fit the pattern for your racial population?

(2) Do you think these four characteristics provide sufficient data for the grouping of individuals into one of the four large racial groups?

2. skin color

a. What is responsible for skin color?

EXERCISE 7. Microscopic examination of skin cross section

Examine the microscope slides of human skin, light and dark. Sketch your observations, highlighting what makes the skin "color" different. What apparently accounts for the "color" difference?

b. How is skin color inherited?

EXERCISE 8. The chemistry of skin color

Based on the information in Figure 9, answer the question that follows it.

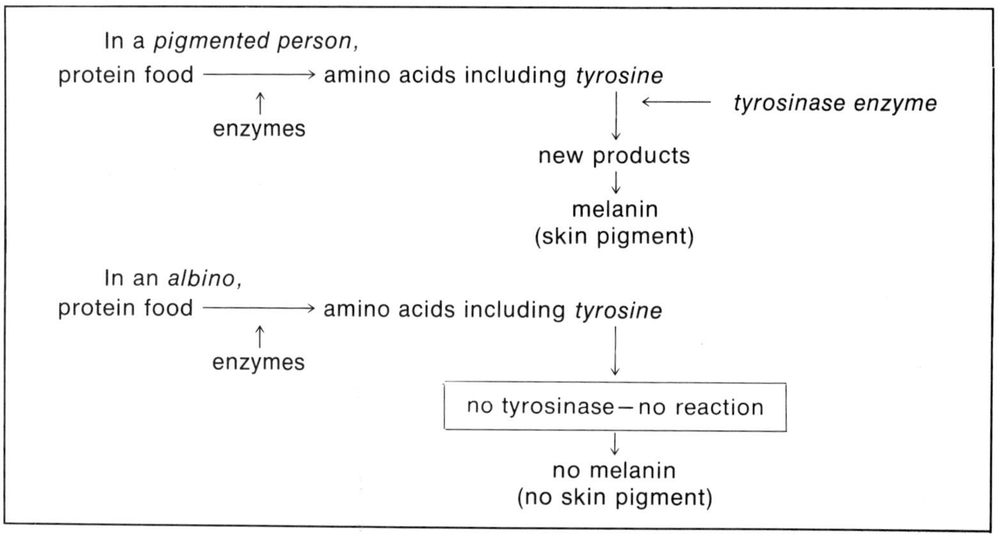

Figure 9. Enzyme control of the production of skin pigment.

How does protein synthesis in an albino person differ from that of a pigmented person?

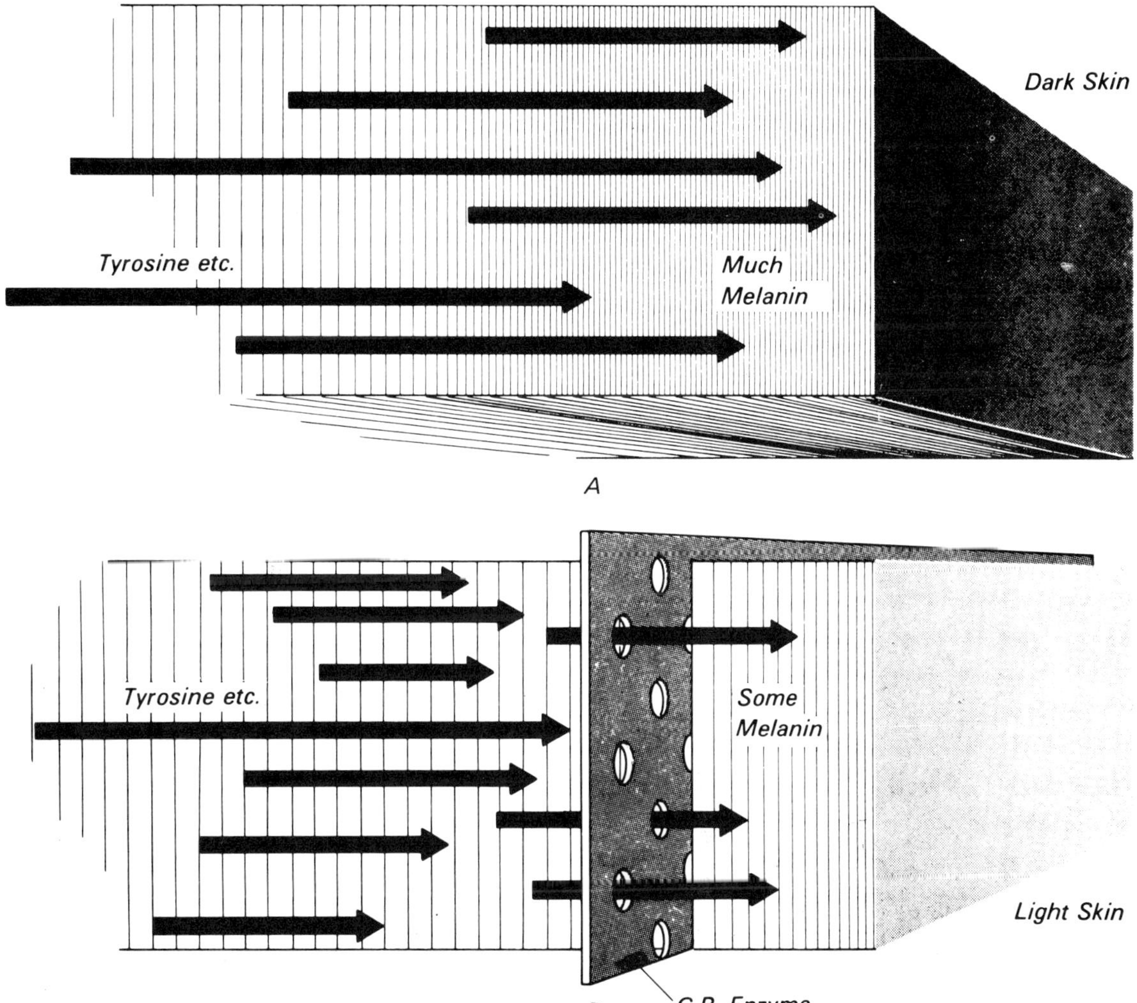

Figure 10. Action of the G. R. enzyme in producing skin pigmentation. *A,* When G. R. enzyme is absent, melanin production proceeds. *B,* Melanin production is slowed when G. R. enzyme is present.

A. Frog cell with Melanin clustered around nucleus. (Light Skin)

B. Same frog cell after Melanin has spread out into the branch parts of the cell. (Darker Skin)

Figure 11. The effect of melanin-stimulating hormone (MSH) on skin color.

C. What Is the Role of Skin Color in Racial Differentiation?

EXERCISE 9. The relation of vitamin D to skin color

Vitamin D is important because it maintains the proper calcium balance in the body. A deficiency of vitamin D (thereby causing a deficiency of calcium deposits) can lead to rickets. Too much vitamin D (thereby causing a surplus of calcium deposits) can lead to brittle bones.

The quantity of vitamin D produced naturally in man is controlled by the amount of sunlight that strikes the cells below the melanin pigment. The more melanin in the skin, the smaller the amount of sunlight that can penetrate to strike the cells below the melanin pigment.

(1) What would you predict is the relationship between skin color and vitamin D production?

(2) What relationship might you suspect exists between the skin color and the location of racial populations in the world?

(3) Study the map in Figure 12. Does the information presented tend to support or reject a skin color/geographical relationship?

(4) The map illustrates the distribution of skin color prior to 1492. Why do you suppose this date was chosen for the map?

(5) In recent years migrations have been widespread. Describe the problems you might expect a dark-skinned person to have in northern or southern latitudes, and the problems you might expect a light-skinned person to have in equatorial regions.

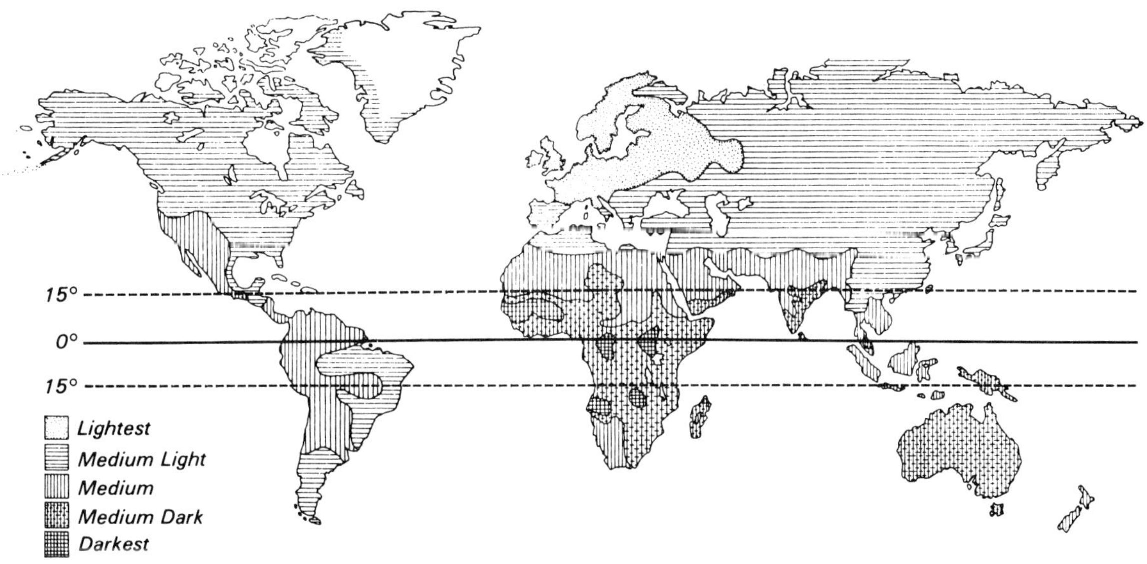

Figure 12. Distribution of skin color prior to 1492. (Reprinted with permission of Macmillan Publishing Company, Inc., from *Man's Evolution* by M. F. Ashley Montague and C. Loring Brace. Copyright © 1965 by M. F. Ashley Montague and C. Loring Brace.)

D. What Are Some Other Racial Characteristics Related to Geography?

Figure 13. Greenland Eskimos living in a region of extreme cold. (Photograph courtesy Peabody Museum, Harvard University.)

Figure 14. Zulu mother and child living in a region of extreme heat and humidity. (Photograph courtesy Peabody Museum, Harvard University.)

EXERCISE 10. Body structure and latitude

Study the individuals in Figures 13 and 14. Consider particularly the ratio of body surface area to body mass.

Why wouldn't it be as efficient for the two populations to exchange latitudes and living conditions? (Review Exercise 4 if you wish.)

E. Do Relationships Exist Between Race and Disease?

Figure 15. A map of the distribution of malaria in the old world. Malaria, a disease caused by the microscopic parasite *Plasmodium falciparum* and its relatives, is the number one infectious killer of man. (Reprinted with permission of Macmillan Publishing Company, Inc., from *Race and Races* by R. A. Goldsby. Copyright © 1971 by Richard A. Goldsby. Adapted from Allison, 1957.)

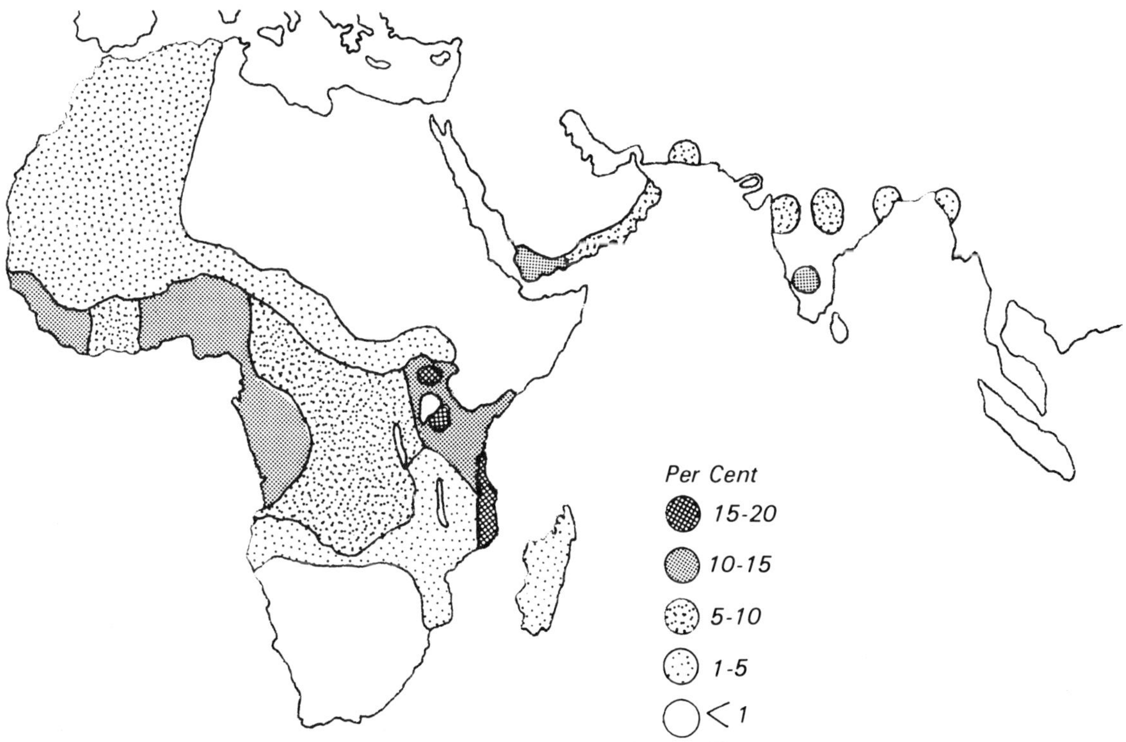

Figure 16. Distribution of sickle-cell anemia, a severe genetic disease. The heterozygous individual with the disease has protection against malaria; the malarial parasite's development is so retarded that the infected individual gets only a very light case of this often fatal disease. This gives the carrier a selective advantage over the individual without the gene. The recessive homozygous individual very rarely lives to reproductive age. (Reprinted with permission of Macmillan Publishing Company, Inc., from *Race and Races* by R. A. Goldsby. Copyright © 1971 by Richard A. Goldsby. Adapted from Allison, 1957.)

EXERCISE 11. What is the relationship between malaria and sickle-cell anemia?

Study the maps in Figures 15 and 16 and answer the following questions.

(1) What is the geographical relationship of sickle-cell anemia to malaria?

(2) If being a carrier of sickle-cell anemia is advantageous, then why doesn't sickle-cell anemia increase to the point of affecting the entire population?

(3) What would you predict would happen to the frequency of the genes for sickle-cell anemia if a population with sickle-cell genes is removed from a malarial area. Explain. (For example: American Blacks.)

IV. ARE RACES OF THE WORLD UNDERGOING CHANGE TODAY?

A. What Is the Effect of Migration on Racial Change?

1. populations that are the result of significant hybridization of two or more formerly different populations
   a. Spanish colonists + American Indians → Spanish Americans
   b. Europeans + African Blacks → American Blacks
2. the Duffy factor

EXERCISE 12. The Duffy factor as an indicator of population hybridization.

The Duffy factor is a blood factor fairly common among Europeans but virtually absent in African Black populations. (The Duffy factor has no known medical significance, therefore no known selective value.)

The original United States Black population came primarily from Western Africa. The original United States White population was primarily European.

Study the data in Table 8 showing the frequency of the Duffy factor in African Black, American White, and American Black populations and answer the questions following it.

TABLE 8.

Duffy frequencies in selected populations.*

POPULATION	FREQUENCY OF DUFFY BLOOD GROUP ALLELE (Fy^a)
African Black population from all slave areas	0.03
Selected West African and Congo populations (principal slave areas):	
Liberia	0.00
Upper Volta	0.00
Dahomey	0.00
Ghana	0.00
Nigeria	0.00
Congo	0.04
American Whites (European ancestors)	0.43
American Blacks	0.08
Regional variations in American Black populations:	
New York City	0.08
Detroit	0.11
Oakland, California	0.09
Charleston, South Carolina	0.02
Evans and Bullock Counties, Georgia	0.05

*Data from Caucasian Genes in American Negroes, T. E. Reed, *Science* Vol. 165, p. 762, 22 August 1969. Copyright 1969 by the American Association for the Advancement of Science.

(1) What evidence in Table 8 supports the idea that American Blacks are a partly hybridized population?

(2) If this trend in hybridization continues for many generations, what changes might be expected in skin color in the United States?

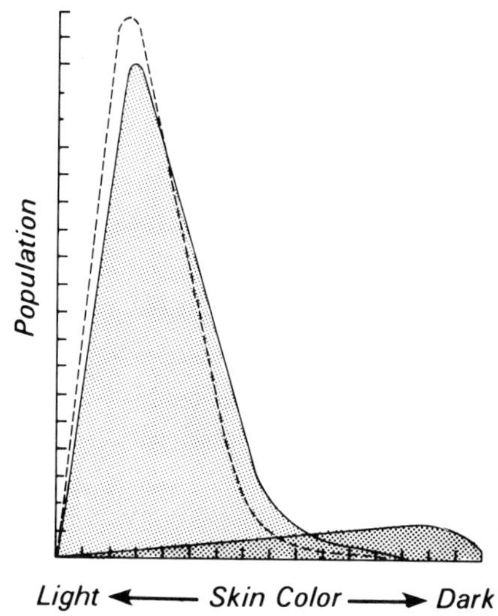

Figure 17. Present and projected distribution of skin color in the American population. (Redrawn from The Biology of the Negro, by Curt Stern. Copyright © 1954 by Scientific American, Inc. All rights reserved.)

B. What Effect Will Man's Continued Mobility Have on the Establishment of New Breeding Populations with Characteristic Gene Frequencies?

1. Israel is currently made up of formerly isolated populations of Jews from Europe, the Middle East, Africa, and Asia Minor.
2. Hawaii is presently undergoing hybridization among Mongoloid, Caucasoid, and Polynesian populations.

V. HOW USEFUL IS THE CONCEPT OF RACE?

A. Biologically, a Race is a Breeding Population.

B. Socially?

The Social Meaning of Human Variation

The present descendants of the seventeenth- and eighteenth-century immigrants into New England probably suffer as little from racial discrimination as any group in the United States. Yet it is impossible to delimit the group on the basis of phenotype. The variability is enormous. The hair may be tow, yellow, sandy, brown, or black. It may be fine or coarse, straight, wavy, or curly. Beard may be heavy or sparse, tough or light. Body build may vary from the Don Quixote to the Sancho Panza type. Eyes may be blue, gray, hazel, brown. One could go on and on listing the phenotypic differences. Within the group, however, there is no consistent discrimination against those members having some specific character or group of characters. There may be some de-

formed or pathologically unattractive individuals who, as a consequence, find themselves at a disadvantage; but these individuals are not regarded as a group to be kept within bounds. Within the established cultural group, human variation is taken as an inevitable part of the nature of things.

It is usually when phenotypic difference becomes identified with a difference of cultural origin that it is likely to give rise to discrimination and become a divisive influence in human society.

APPENDIX I

Answers to the statements on Page 1 concerning the concept of race:

1. Possibly true but speculative. It has not been established whether human beings evolved in hot equatorial climates or in subtropical or warmer temperate climates. It has also not been established what the skin color of early humans was; fossil skin has not been found with fossil human bones.
2. True
3. False
4. True
5. False
6. False
7. True
8. False
9. False
10. This statement is difficult to answer because of the degree of interbreeding that has occurred, particularly between Blacks of African ancestry and Whites of European ancestry.
11. False
12. False
13. False
14. False
15. False

APPENDIX II

Answers to Exercises

EXERCISE 1.

(1) small population size, isolation, chance

(2) Because of greater density and greater mobility of human populations today, such changes would either not occur or would take a much longer time to occur. Certain isolated tribes still continue to exist, however, and in these tribes random genetic drift is easily possible.

ADDITIONAL INQUIRIES

1. Research the direct effects of the sun's radiation on the production of melanin and the occurrence of sun-burned or sun-tanned skin.
2. Additional reading:

   Osborne, Richard H. 1971. The biological and social meaning of race. W. H. Freeman and Company, San Francisco, California.

GLOSSARY

ALBINO: An organism lacking a gene necessary for the synthesis of pigment factors.

BAIB: A chemical excreted in the urine of some individuals. The chemical has no known value to man.

DUFFY FACTOR: A blood factor fairly common among Europeans but nearly absent in African Black populations. It is a genetic factor with no known medical significance.

GENE POOL: Sum of all the genes in an interbreeding population.

GENETIC DRIFT: Changes in frequencies of genes in the gene pool as a result of chance happenings; usually the characteristics of the genes involved confer no advantage or disadvantage on the population having the genes in its gene pool.

ISOLATION: The separation of two populations from each other; geographical distance or cultural barriers may be a cause.

MELANIN: A dark pigment formed in the body and deposited in different amounts in the skin, hair, and nerve tissues.

POLYTYPIC SPECIES: A species that has two or more subspecies or subcategories.

RACE: A breeding population of individuals, identifiable by the frequency with which a number of inherited traits appear in that population.

SPECIES: A group of actually or potentially interbreeding natural populations that is reproductively isolated from other such groups.

REFERENCES

Brace, C. L., and M. F. Montagu. 1965. Man's evolution: an introduction to physical anthropology. The Macmillan Company, New York. 352 pp.

Cohen, R. 1958. The color of man. Random House, Inc., New York. 114 pp.

Coon, C. S. 1965. The living races of man. Alfred A. Knopf, New York. 344 pp.

Dobzhansky, T. 1962. Mankind evolving. Yale University Press, New Haven, Connecticut. 381 pp.

Goldsby, R. A. 1971. Race and races. The Macmillan Company, New York. 132 pp.

King, J. C. 1971. The biology of race. Harcourt Brace Jovanovich, Inc., New York. 189 pp.

Lerner, I. M. 1968. Heredity, evolution, and society. W. H. Freeman and Company, San Francisco, California.

Matsunaga, E. 1962. The dimorphism in human normal cerumen. Annals of Human Genetics. *25*(4):277–286.

Reed, T. E. 1969. Caucasian genes in American Negroes. Science. *165*(3895):762–768.

Schreider, E. 1957. Ecological rules and body-heat regulation in man. Nature. *179*:915–916.

Schreider, E. 1957. Gradients écologiques régulation thermique et différénciation humaine. Biotypologie. *18*:168–183.

Schwidetzky, I. 1962. Die neue Rassenkunde. Gustav Fischer Verlag, Stuttgart.

Stern, C. 1954. The biology of the negro. Scientific American. *191*(4):80–87.

Woolf, C. M., and F. C. Dukepoo. 1969. Hopi Indians, inbreeding and albinism. Science. *164*:30–37.

study guide

the problem of population

minicourse

the populations of two space ships

These two spaceships are very similar. They both float through endless space. They both have very delicately balanced self-contained systems which support life. Each creates and absorbs oxygen and wastes, but each life-support system is fragile and easily can be destroyed. The greatest difference is that the Apollo space ship still had only three men aboard at the end of its ten-day trip. In the same ten days, however, the earth's population grew by two million persons. . .

Prepared by John R. Bermingham, State Senator, Denver, Colorado

PREREQUISITES: There are no prerequisites for this minicourse.

I. DOES A POPULATION PROBLEM EXIST?

A. Survey of Opinion

TABLE 1.

Time of interview: 1959. (After The Gallup Poll.)

Have you heard or read about the great increase in population that is predicted for the world during the next few decades?	
Yes	75%
No	25%
Asked of those who applied in the affirmative: Are you worried or not worried about this population increase?	
Worried	21%
Not Worried	69%
No Opinion	10%

TABLE 2..

Time of interview: 1971. (After The Gallup Poll.)

How serious a problem for the country do you think present population growth in the United States is?	
Present U.S. population growth is a major problem now	41%
Not a problem now, but likely to be a problem by the year 2000	46%
Not a problem now, and don't expect it to become problem; or don't know	13%

B. Population Estimates

EXERCISE 1. What are some population data?

How well do your estimates of population agree with recent United Nations and U.S. Census data?

(1) The estimated world population in 1973 was __________ people.

(2) The estimated U.S. population in 1973 was __________ people.

(3) At the time of the most recent U.S. census, the population of the state in which I live was __________ people.

(4) How many states showed a population loss at the last census? __________ Name one __________________.

(5) By the year 2100 the world population will be __________ people.

C. Population Facts to Consider

1. The world population is approaching 4 billion persons, increasing by approximately 75 million each year.
2. At the current rate of increase (2 percent), every three years the world population grows by a number greater than the present U.S. population.
3. One million people are added to the world population every five days.
4. Five percent of all the people who have ever lived on earth are living today.
5. The great tidal wave that struck East Pakistan in 1970 killed an estimated half million people.
6. Developed countries consume a disproportionate share of global fuel, mineral, and food resources, while polluting air and water at a significantly higher rate.

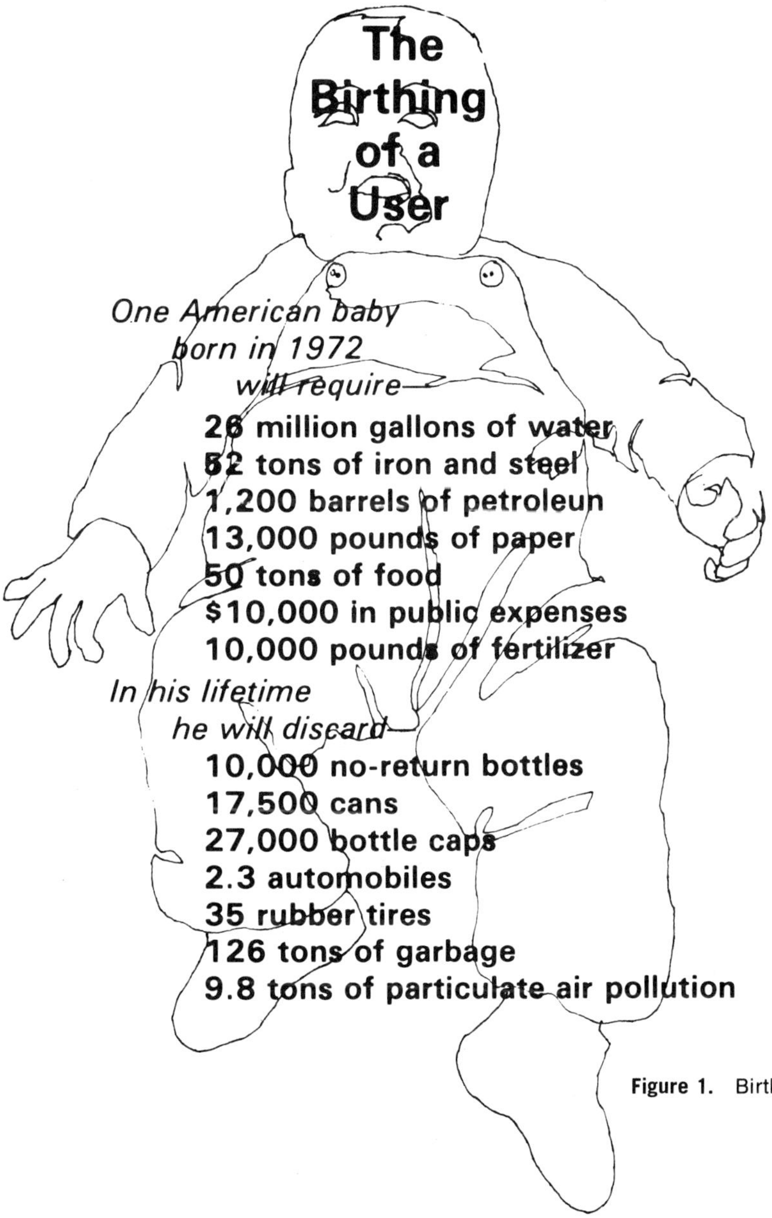

Figure 1. Birth of a user. (From Meadows. 1972.)

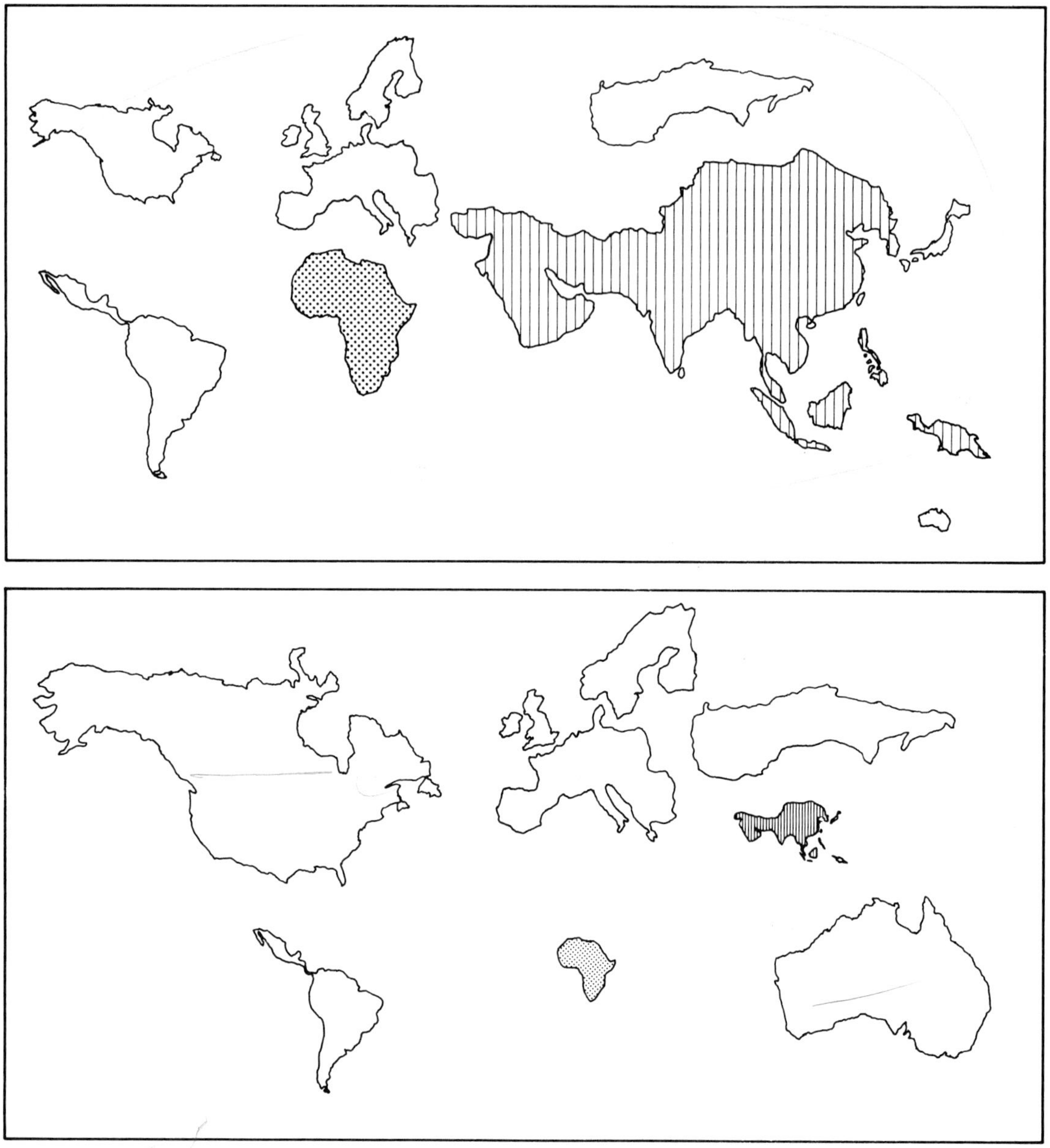

Figure 2. World maps scaled to population (*above*) and per capita Gross National Product (*below*). (From Population Reference Bureau Press Release, Population Reference Bureau, Inc., Washington, D.C.)

II. HOW AND WHY HAS THE HUMAN POPULATION GROWN IN NUMBERS?

A. Population Growth Curves

Figure 3. Dandelion.

Figure 4. Adult frog with egg mass. (Photo courtesy Lynwood M. Chace from National Audubon Society.)

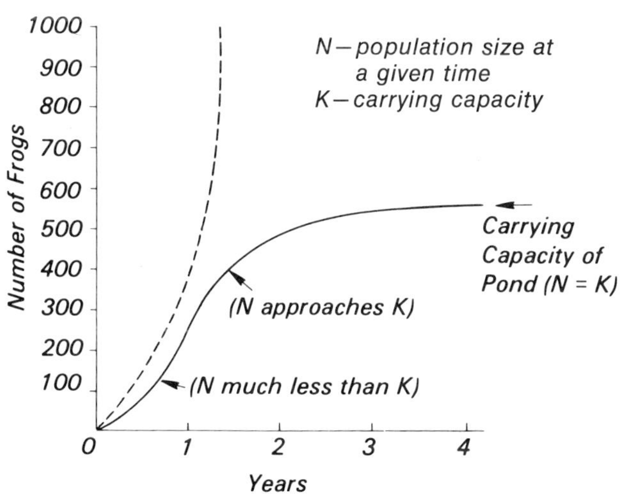

Figure 5. Hypothetical case of growth of a frog population in a pond, starting with one pair. Dotted line is theoretical maximum increase in population resulting from reproductive potential. Note its logarithmic nature. Solid line is S-shaped curve produced by environmental resistance. The population eventually reaches an equilibrium level, called the "carrying capacity of the environment." (From Clark. 1973.)

(By permission of John Hart and Field Enterprises, Inc.)

Man is part of a natural system, the earth, and is ultimately subject to the limits of the system.

B. Do Populations Increase at a Constant Rate?

EXERCISE 2. How does the size of a yeast population increase over a period of time?

In order to allow you the time to complete this exercise, the yeast populations have already been grown. Each culture tube at the central study area has a number and label recording the time in hours the population was left to grow before being "fixed" with a 4 percent formalin solution.

You will not count the actual number of cells, but rather analyze the turbidity or cloudiness of the solution with a spectrophotometer. This is related to the number of cells present.

(1) Measure the absorbency (or optical density) of each tube of yeast population and record it in Table 3.

TABLE 3.

Student data table.

TUBE #	TIME (Hours)	OPTICAL DENSITY	TUBE #	TIME (Hours)	OPTICAL DENSITY

(2) Plot the values of absorption versus time in hours, and connect the points with a continuous line on the graph in Figure 7 on Page 6. Determine the best scale to use for each axis before you begin your graph.

(3) Is the increase a constant rate?

(4) Notice that the resulting graph has three distinct phases. Can you explain why these phases might occur?

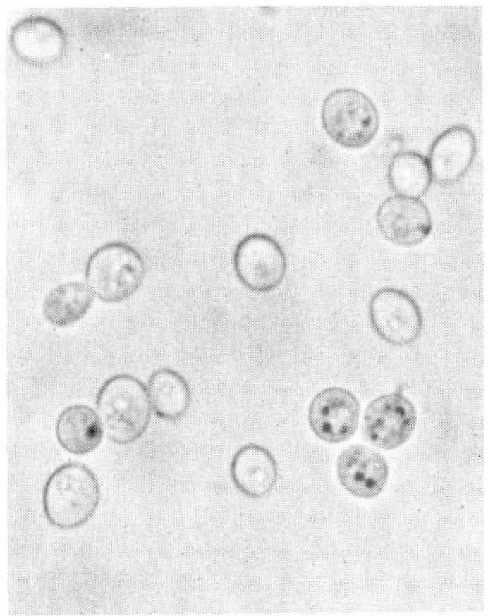

Figure 6. Budding and reproduction by the yeast *Saccharomyces cerevisiae.* (Photograph courtesy Victor Larsen.)

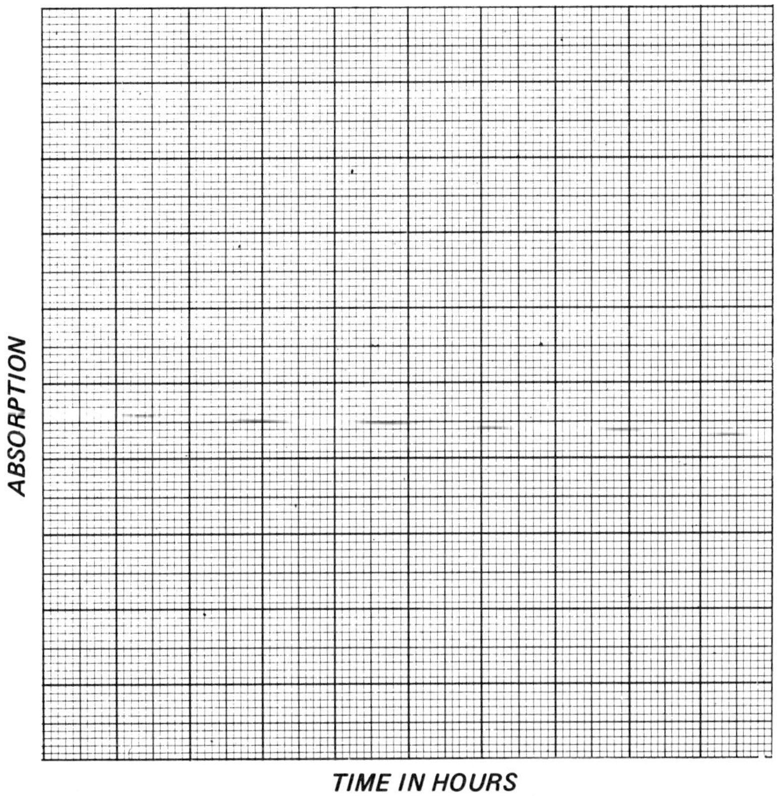

Figure 7. Graph for data obtained in Exercise 2.

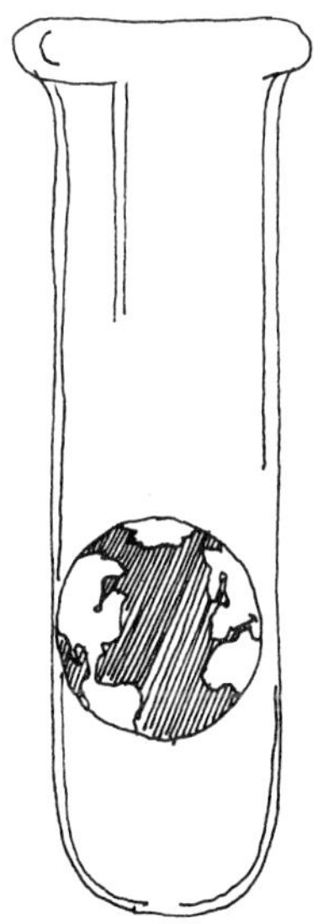

EXERCISE 3. Estimating human population on growth curve

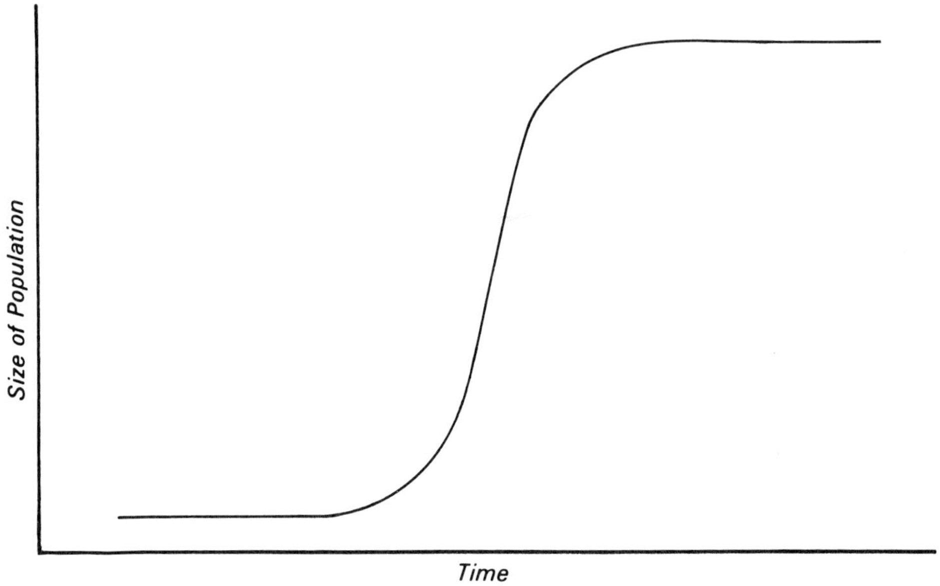

Figure 8. Hypothetical growth curve.

(1) Place an X on the growth curve at the point you feel represents the human population today.

(2) Place an X where the population stood 200 years ago, when the United States first became a country.

(3) How many years will pass before the population will begin to level off?

(4) What factors will be responsible for the leveling off?

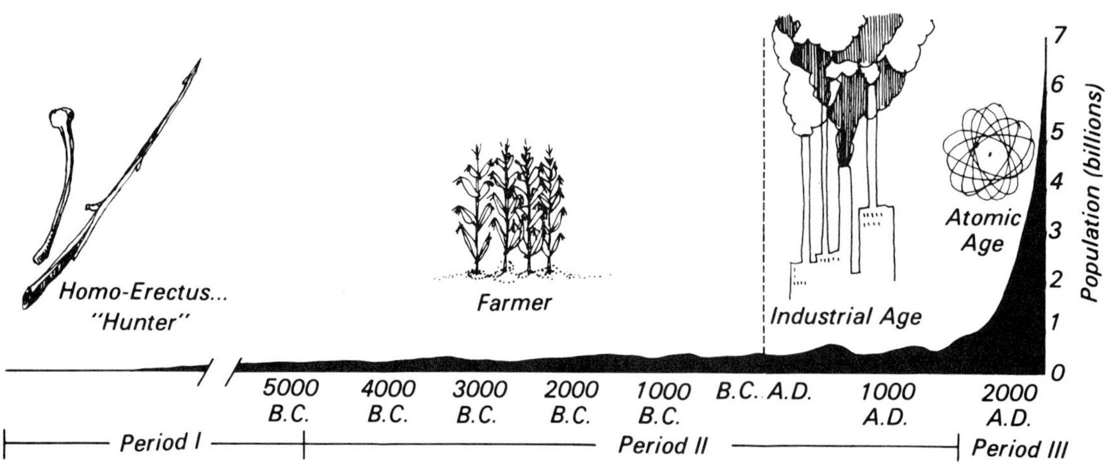

Figure 9. Historical growth of the human population.

III. WHAT STAGES CHARACTERIZE THE DEMOGRAPHIC TRANSITION OF HUMAN POPULATION?

A. Demographic Definitions

1. *Demography*—The statistical study of human populations.
2. *Birth Rate*—Live births per 1000 population.
3. *Death Rate*—Deaths per 1000 population.
4. *Infant Death Rate*—Deaths per 1000 babies under the age of one year.

TABLE 4.

Population data for several countries (1971).*

	Population estimates mid-1969 (Millions)	Birth rate per 1000 population	Death rate per 1000 population	Current rate of population growth (MAGR)	Number of years to double population	Population under 15 years (percent)
World	3706	34	14	2.0	35	37
United States	207	18.2	9.3	1.1	58	30
Denmark	5.0	14.6	9.8	0.5	88	24
East Germany	16.2	14.0	14.3	0.1	233	22
Sweden	8.1	13.5	10.4	0.5	88	21
Colombia	22.1	44.0	11.0	3.4	21	47
Ecuador	6.3	45.0	11.0	3.4	21	48
Venezuela	11.1	41.0	8.0	3.4	21	46
Kenya	11.2	50.0	20.0	3.1	23	46
Uganda	8.8	43.0	18.0	2.6	27	41
Nigeria	56.5	50.0	25.0	2.6	27	43
India	570	42.0	17.0	2.6	27	41
Pakistan	142	50.0	18.0	3.3	21	45
Indonesia	125	47.0	19.0	2.9	24	42

*From *Biology* by James D. Ebert, Ariel G. Loewy, Richard S. Miller, and Howard A. Schneiderman.

TABLE 5.

Infant mortality rates in several countries for both 1901–1905 and 1973.*

INFANT MORTALITY RATES					
COUNTRY	**1901–1905**	**1973**	**COUNTRY**	**1901–1905**	**1973**
Sweden	91	11	Belgium	148	20
Finland	131	11	Ireland	94	20
Netherlands	136	11	Germany (Federal Republic of)	199	23
Norway	80	13	Spain	172	28
Japan	– –	13	Italy	167	28
France	139	13	Ceylon (Sri Lanka)	171	48
Denmark	119	14	Portugal	144	50
Switzerland	134	14	Mexico	132 (1928–1938)	69
Australia	97	17	Chile	264	88
Canada	98 (1921–1925)	18	Egypt	– –	118
United Kingdom	138	18	India	215	139
United States	74 (1921–1925)	18			

*Data for 1901–1905 from *Demographic Year Book,* 1952, United Nations. Data for 1973 from *Population and Vital Statistics Report,* Series A, Volume 25, Number 2, 1973, United Nations.

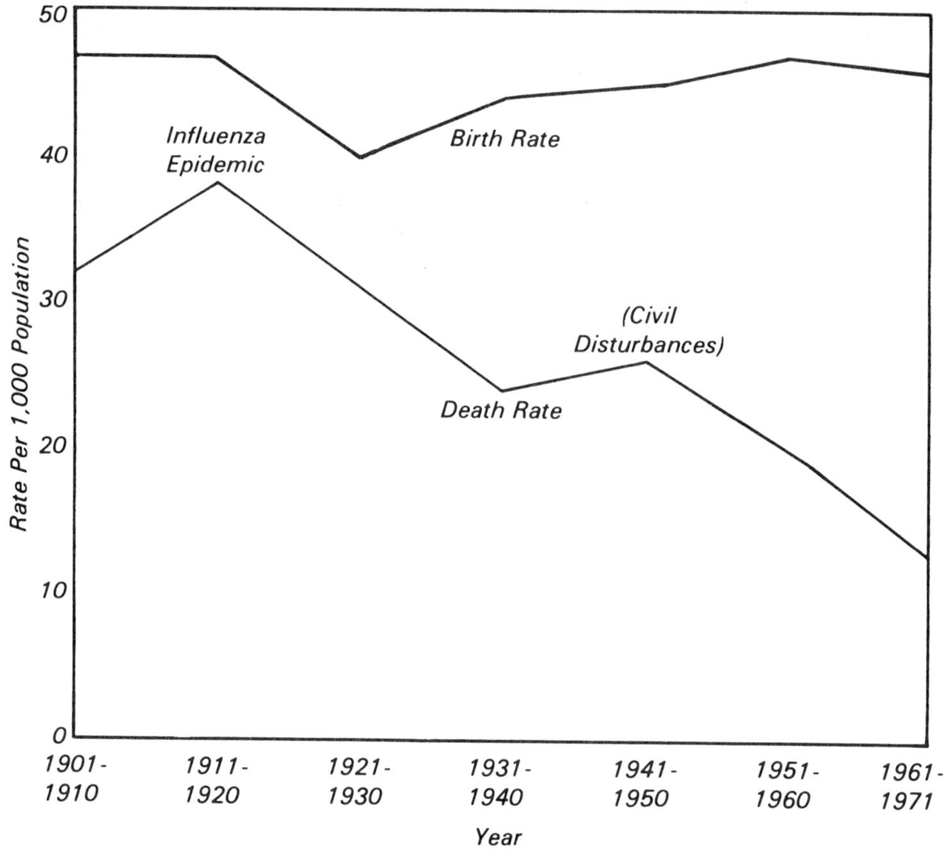

Figure 10. Average birth and death rates, Punjab Province, Pakistan, 1901–1971. (From Population Bulletin Vol. 29, No. 4. *Pakistan: A Demographic Report,* Population Reference Bureau, Inc., Washington, D.C.)

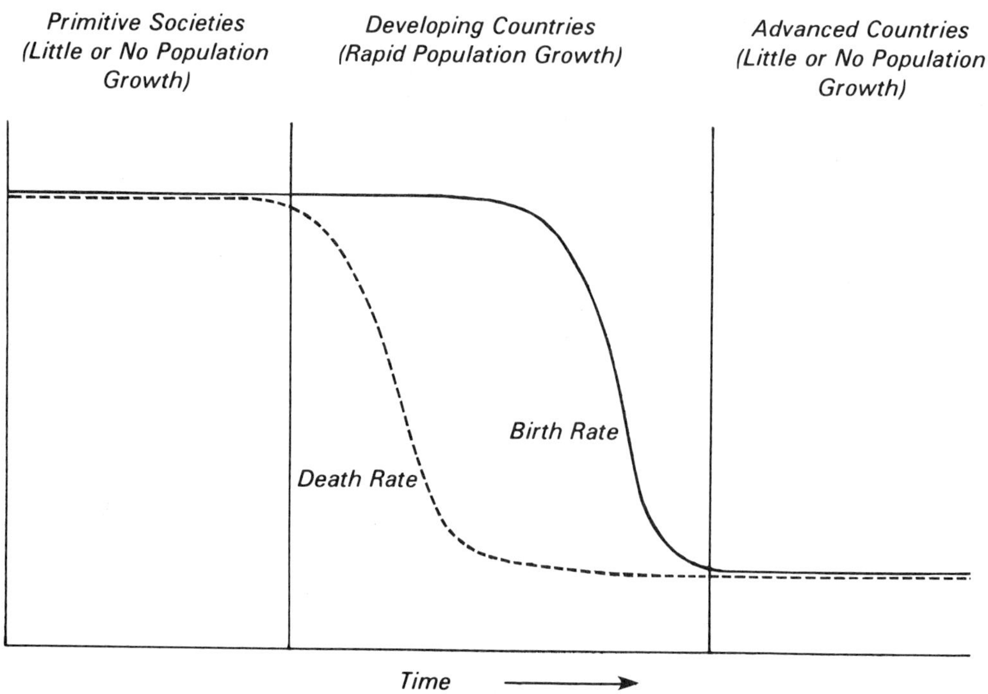

Figure 11. Demographic shift through which countries pass in birth and death rates before attaining population stability. (From Wilson, et al. 1973.)

B. Characteristics of Transitional Stages

1st Stage – high birth rates and high death rates
– per capita income low
– tradition supports high birth rate (large families offer practical economic advantages)
– high birth and death rates balance each other; slow population growth

2nd Stage – high birth rates and lower death rates
– increased food production and economic development
– better medical practices contribute to lower death rate
– falling death rates with continuing high birth rates give rise to rapid population growth

3rd Stage – low birth rates and low death rates
– people deliberately control birth rate
– families are smaller, per capita income higher
– industrialization, urbanization, education seems to influence people's decisions regarding birth rate
– population grows at an economically manageable rate or levels off altogether

The above three stages compose the *demographic transition.*

IV. WHAT IS THE MAJOR FACTOR INFLUENCING POTENTIAL GROWTH OF THE WORLD'S POPULATION?

A. Age Classes Expressed as Age Pyramids

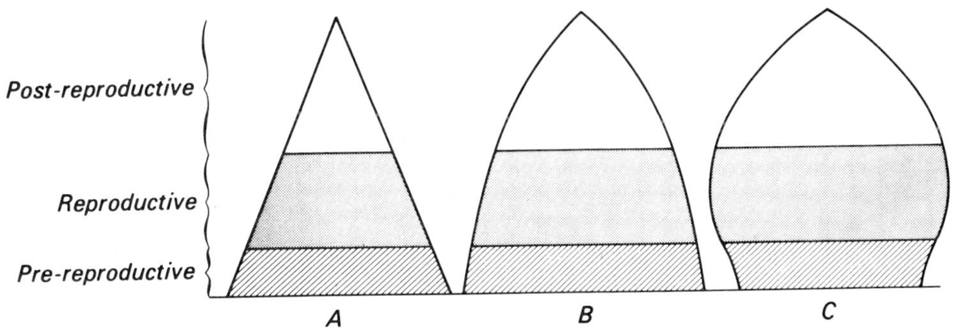

Figure 12. Age pyramids for three types of populations. *A,* Rapidly growing; *B,* stable; *C,* declining. (From Clark. 1973. After Edward J. Kormondy, *Concepts of Ecology* © 1969. Reprinted by permission of Prentice-Hall, Inc., Englewood Cliffs, New Jersey.)

B. Inferences That Can Be Drawn from the Structure of Population Pyramids

EXERCISE 4.

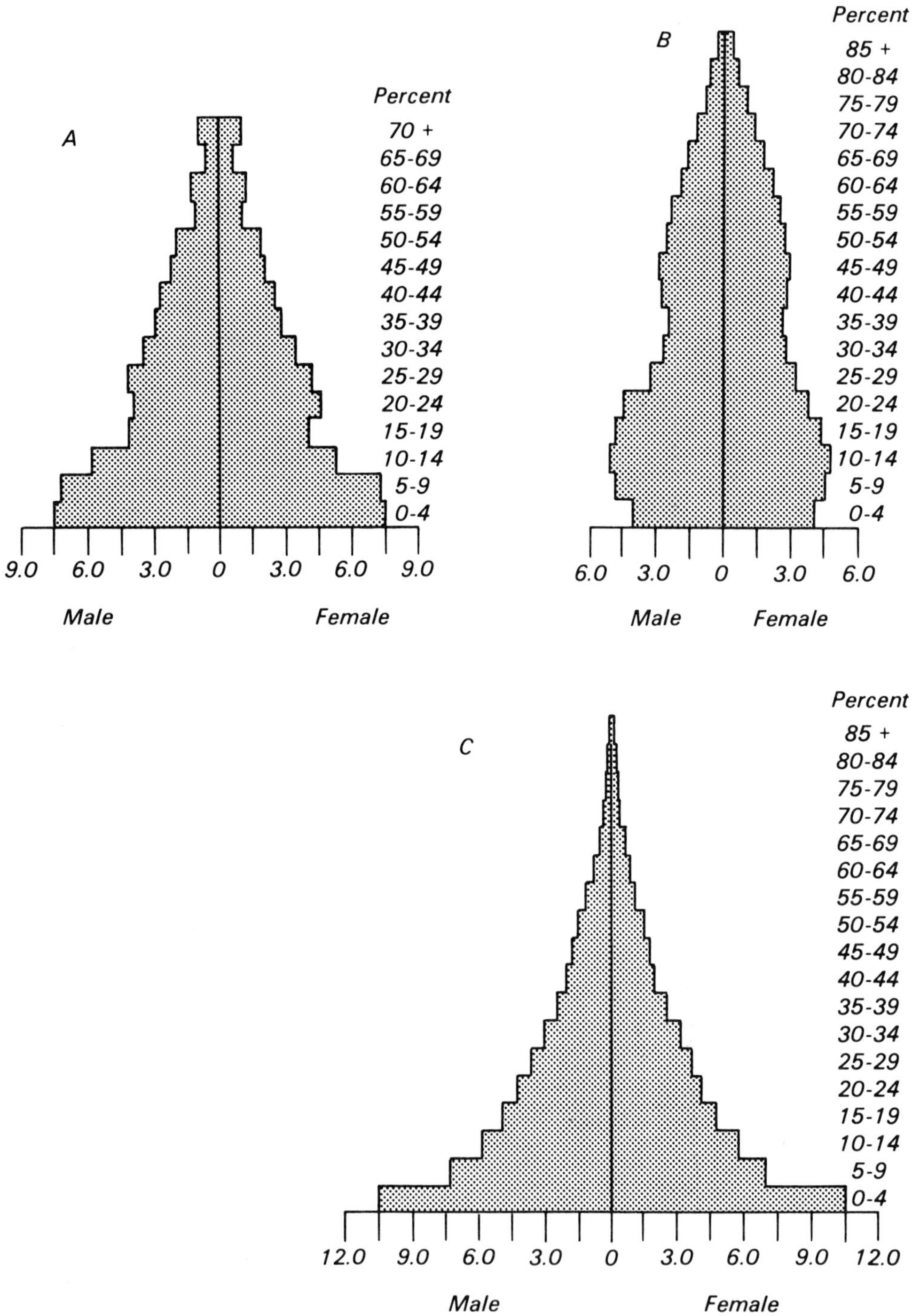

Figure 13. Sample population pyramids. (After *Principles of Demography*, Donald J. Bogue. Copyright © 1969 by John Wiley and Sons, Inc. Reprinted by permission of John Wiley and Sons, Inc.)

After observing the three population pyramids in Figure 13, answer the following questions:

(1) How are the three pyramids similar? How are they different?

(2) To what factors may the differences be attributed?

(3) Which pyramid do you think most represents a highly developed, industrialized nation?

(4) Which pyramid represents a developing, agricultural country?

C. What Do Pyramids Tell Us About the Stages of Population Growth?

EXERCISE 5.

Based on the information presented in the six population pyramids in Figure 14, answer the following questions:

(1) Which two countries appear to have rapidly expanding populations? Why?

(2) Which two countries appear to have relatively stable populations? Why?

(3) Which country appears to have most recently begun a trend toward a diminishing population? Why?

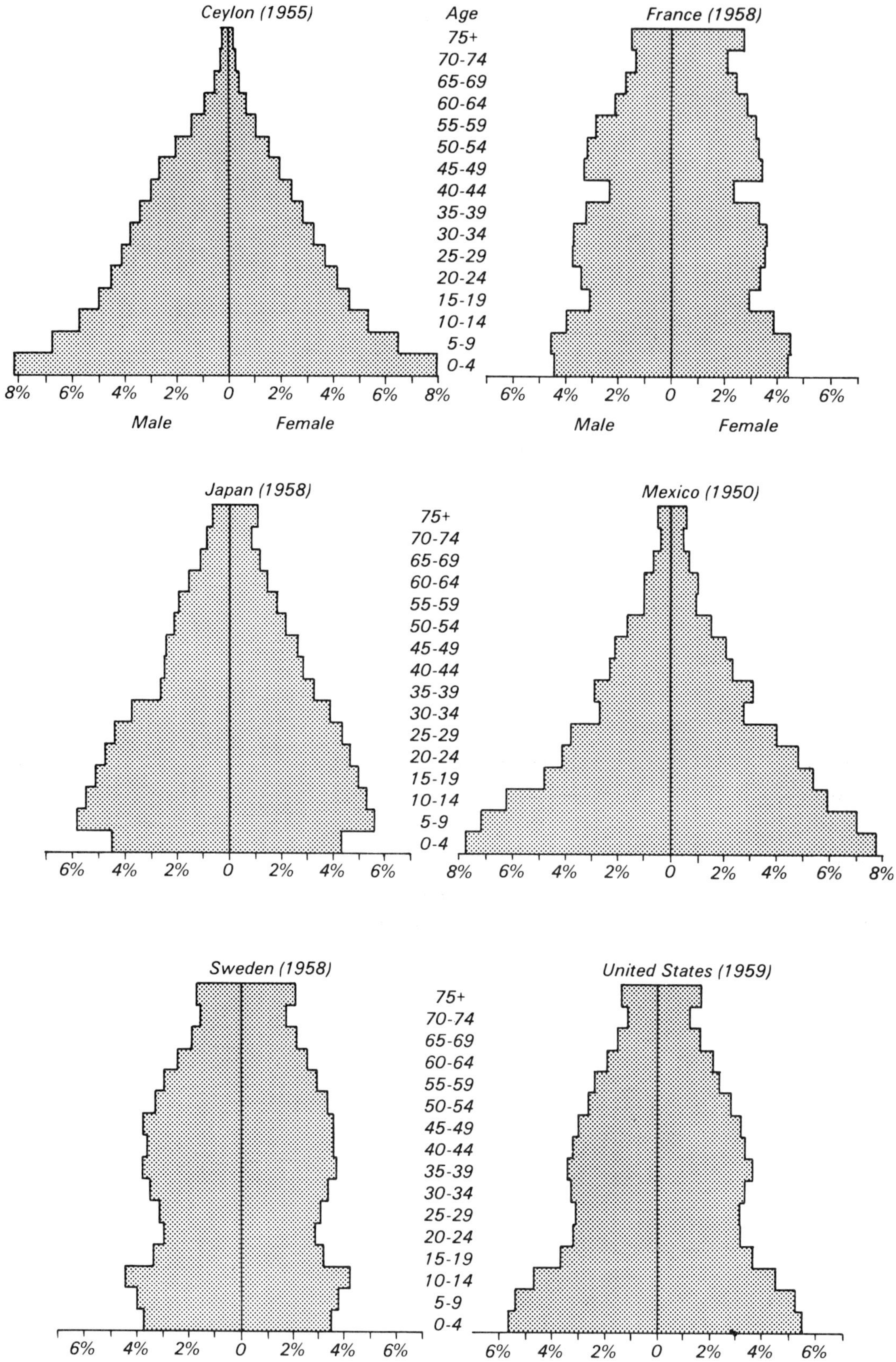

Figure 14. Age-sex pyramids of several countries, 1950–1959. (From Population Bulletin Vol. 16, No. 8, 1960, Population Reference Bureau, Inc., Washington, D.C.)

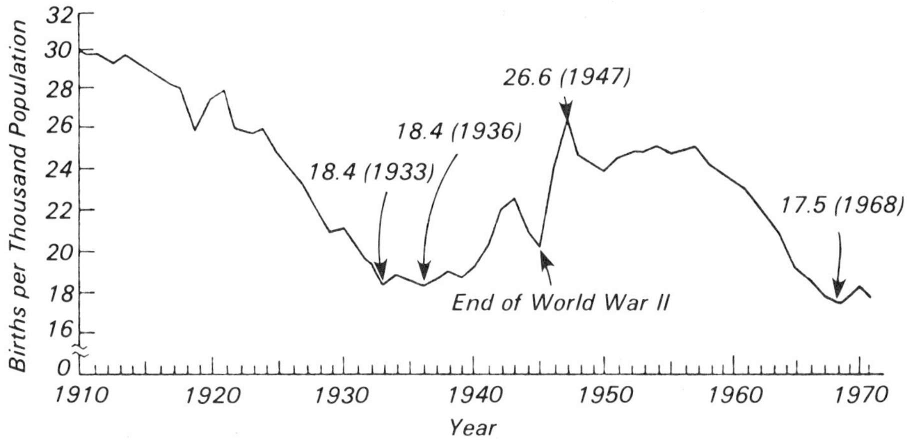

Figure 15. Birth rate in United States, 1910–1971. (After *Population Profile*, Population Reference Bureau, Inc., Washington, D.C., March 1967.)

Fertility Rate: Number of live births per 1000 females 15 to 44 years of age.

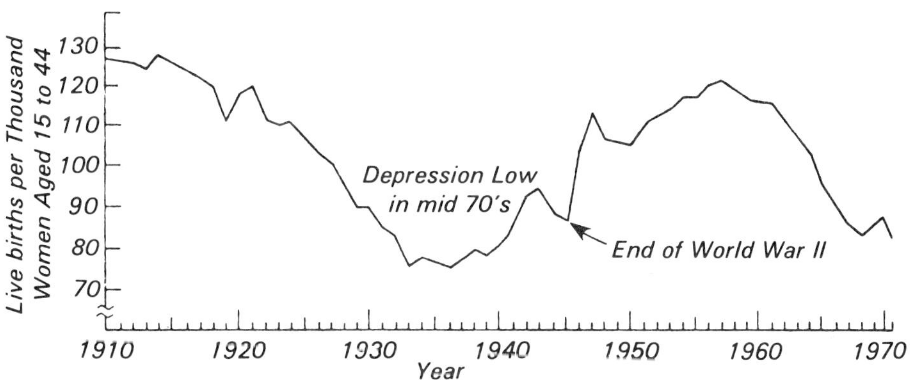

Figure 16. Fertility rate in the United States, 1910–1971. (After *Population Profile*, Population Reference Bureau, Inc., Washington, D.C., March 1967.)

V. WHAT OPTIONS DOES POPULATION GROWTH HOLD FOR THE WORLD?

A. The Goal—To Achieve a Net Reproductive Rate of 1

1. A *net reproductive rate* of 1 indicates equality between fertility and mortality.
2. Fertility resulting in a net reproductive rate of 1 is called *replacement fertility.*
3. *Total fertility rate* is the average number of live children a female would bear during her years of potential motherhood.
4. In the United States (and some other developed countries) a total fertility rate of about 2.1 is the *replacement fertility.*

B. What Are the Future Projections?

PROJECTION 5: *Assumes the present fertility level declines to NRR of 1 in 70 years*

PROJECTION 4: *Assumes NRR = 1 in 50 years*

PROJECTION 3: *Assumes NRR = 1 in 30 years*

PROJECTION 2: *Assumes NRR = 1 in 10 years*

PROJECTION 1: *Assumes NRR = 1 immediately*

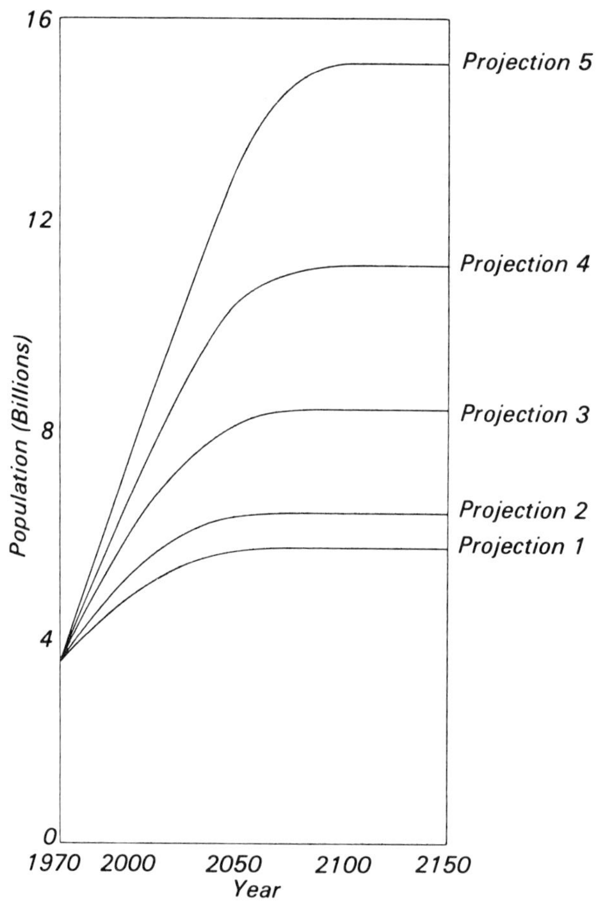

Figure 17. World population projections based on five assumptions when a net reproductive rate (NRR) of 1 might be achieved and maintained. The projections assume that the present fertility level declines to a level corresponding to a NRR of 1 immediately (Projection 1) or over a period of 10, 30, 50, and 70 years (Projections 2 to 5). (From The Prospects for a Stationary World Population, by Tomas Frejka. Copyright © 1973 by Scientific American, Inc. All rights reserved.)

EXERCISE 6. Possible future populations

Use the projected growth curves in Figure 17 to answer the following questions:

(1) Projection 3 assumes the total fertility rate will drop to 2.2 (considered replacement) by the year 2000 (30 years after 1970). What will the population already have grown to by the year 2000? What will it be by the time the population levels off in about 2100?

(2) Projection 5 assumes that the total fertility rate will have decreased to only 3.5 by the year 2000. What will the population be by the year 2050? By the year 2100?

(3) What will life be like for a person living in the year 2100 if Projection 3 is achieved? If Projection 5 is achieved?

VI. WHAT ARE THE FUTURE OPTIONS?

A. UNCONTROLLED FUTURE

There will be a continued increase in the world's population, especially in the newer developing nations. The problems of famine, pollution, energy crisis, and war will be the major factors contributing toward population control.

"Although we used a computer for our study of long-term growth trends, in fact the results don't require any understanding of the computer—they're rather straightforward. If we continue these growth trends—with population doubling every thirty years, with energy consumption doubling every fifteen years and so forth—then the result will be, sometime in the near future, that we will grow past *important limits. We will go to such high levels that we can't sustain population anymore, and there'll be a collapse.*

I often ask myself why it is that our generation, out of all those hundreds that have lived on earth, has to be the one to deal with this set of problems. Why is it we who will be alive for the first time in global history that population, material growth, and other trends come to an end? Well, I don't know the answer to that, but I know that because others before have ignored the problems, they finally fall on us to solve."

(Excerpts from an interview with Dennis Meadows, one of the authors of *The Limits of Growth.*)

B. STABILIZED FUTURE

Through concentrated and cooperative efforts the world's population will reach a zero population growth before the "critical limits" of growth are surpassed.

"In the past few years we have had a number of major changes, which have been triggered by acts of analysis. It's true that there's a general change of consciousness. Our change is mediated by TV, it's mediated by higher education, so that . . . we have people everywhere rising up as leaders, forming little groups, protesting . . . trying to change their own communities. Communities have begun to limit their growth—limit their sewers, their schools, their expansion into the countryside, the height of their buildings, the automobiles coming into town—because they've realized that this struck a chord, this explained a lot of things that were happening. The result is, all over the world, one begins to have an ecology consciousness, a limits-to-growth consciousness, *which is world wide and is moving us in the direction of more responsibility for our global future, our global changes, our global interactions. I could imagine now, that in the years ahead, there're several great types of analysis that we should have, of this same sort. For example, we need an analysis of those megafamines that are* likely *to occur in the next few years, killing tens of millions of people, perhaps. We need to bring together scientists, including meteorologists to talk about the weather, demographers, people . . . biologists who work in the green revolution, sociologists, political scientists who can discuss how these groups can cross political boundaries if they're starving in a particular country. And forecasters, to anticipate how the various factors are going to interact. And we need conferences on global management, on management of ocean resources—and one can name just one area after another where an act of analysis will transform our realization of what's happening as we move into these crises ahead, and will make it possible for us to deal with them, instead of giving up in hopeless despair when the next great impact comes."*

(Excerpts from an interview with John Platt, Associate Director of the Mental Health Research Institute, University of Michigan.)

C. EXPANDING FUTURE

Following a period of "runaway" growth there will be a modification of life styles and behavior. People will adjust to new technology, 24-hour-a-day activities, more leisure time, and more crowded environs.

"People who do 'futures research' as I at least try to do—or people who are called 'futurists'—get a reputation of being doomsayers, of having only negative things to say about the world. At least many of them do. Well, I think there is real danger that we may not be pessimistic enough about the magnitude of the task right in front of us . . . because we really have an overall system which month by month, year by year, works less and less well. And you see manifestations of this in the energy crisis, in the food crisis that's coming, and all the various dislocations. You can't get good postal service any more, you can't buy good products any more—you hear all these complaints, and they get louder and louder, and the gas lines get longer and longer . . . so, there is this dilemma that, as the system is now operated—in the interests of increasing efficiency, increasing growth, increasing technological growth and advance, increasing productivity—it grinds up people.

The problem seems to be that in the system's pursuit of growth and productivity, and technological growth, and efficiency, that people end up with their own ends thwarted—that in the pursuit of efficiency, we displace people from jobs, or we give . . . we end up with people having relatively meaningless jobs; in the pursuit of productivity we use up the earth's resources and we despoil the environment. So, the challenge is to so modify the system, somehow, that it again is successful, in some other terms, perhaps.

I think the second real danger is that we won't be optimistic enough about where we might be in fifteen, or twenty, or twenty-five years, after we go through the agony of a rather major modification of society. Those things are never easy! And we have never in human history had one come as rapidly as this. But nevertheless these are like growing pains, and if we get through, we enter a whole new phase, and there's no reason to assume that that couldn't be a better time than we've ever known."

(Excerpts from an interview with Willis Harman, Director of the Center for the Study of Social Policy, Stanford Research Institute.)

APPENDIX I
Answers to Exercises

EXERCISE 1.

(1) 3860 millions (3.86 billions)*

(2) 210.3 millions*

*From *Total Population Estimates for World, Regions and Countries, Each Year, 1950–1985*, Population Division Working Paper No. 34, United Nations, 1970.

(3) and (4) Population data of the fifty states (only North and South Dakota, the District of Columbia, and West Virginia declined): See Table 6 on Pages 19 and 20.

EXERCISE 6.

(1) 5.9 billion by year 2000
8.4 billion by year 2100

(2) 13 billion by year 2050
15.1 billion by year 2100

(3) ?

ADDITIONAL INQUIRIES

1. Beginning with an initial population of two individuals, construct a hypothetical growth curve on both semi-log paper and regular graph paper. Assume the population increases in a geometric progression (i.e., two, four, eight, sixteen) for 20 generations.

2. Construct growth curves of a real population of fruit flies (*Drosophila melanogaster*) on both semi-log paper and regular graph paper. Begin with an initial population of two males and two virgin females. Record the population count for four generations. Allow from 12 to 14 days for each generation to complete its cycle. Devise your own system for sampling and estimating the population size when it becomes too large to count all the individuals.

3. Discuss in a written report the relationship that exists between population size per unit area and one of the following social problems: crime, hunger, pollution, energy supply, war. State your own personal views and include some supportive data.

4. Additional readings:

Hardin, Garrett. 1974. Living on a lifeboat. BioScience. *24*(10):561–568.

Malthus, Thomas R. 1959. Population: the first essay. University of Michigan Press, Ann Arbor, Michigan.

U.S. News and World Report. 1974. What life will be like 20 years in the future. *76*:72–75.

GLOSSARY

BIRTH RATE: Live births per 1000 population.

CARRYING CAPACITY: The upper level of population that can persist, beyond which no permanent increase can occur because of environmental resistance.

DEMOGRAPHY: The statistical study of human populations, especially with reference to size and density, distribution, and vital statistics.

ENVIRONMENTAL RESISTANCE: All en-

TABLE 6.

Rank of states by population, 1920–1970. From U.S. Bureau Summary, 1970.

STATES	1970 Rank	1970 Population	1960 Rank	1960 Population
Alabama	21	3,444,165	19	3,266,740
Alaska	51	300,382	51	226,167
Arizona	33	1,770,900	35	1,302,161
Arkansas	32	1,923,295	31	1,786,272
California	1	19,953,134	2	15,717,204
Colorado	30	2,207,259	33	1,753,947
Connecticut	24	3,031,709	25	2,535,234
Delaware	47	548,104	47	446,292
District of Columbia	41	756,510	40	763,956
Florida	9	6,789,443	10	4,951,560
Georgia	15	4,589,575	16	3,943,116
Hawaii	40	768,561	44	632,772
Idaho	43	712,567	43	667,191
Illinois	5	11,113,976	4	10,081,158
Indiana	11	5,193,669	11	4,662,498
Iowa	25	2,824,376	24	2,757,537
Kansas	28	2,246,578	28	2,178,611
Kentucky	23	3,218,706	22	3,038,156
Louisiana	20	3,641,306	20	3,257,022
Maine	38	992,048	36	969,265
Maryland	18	3,922,399	21	3,100,689
Massachusetts	10	5,689,170	9	5,148,578
Michigan	7	8,875,083	7	7,823,194
Minnesota	19	3,804,971	18	3,413,864
Mississippi	29	2,216,912	29	2,178,141
Missouri	13	4,676,501	13	4,319,813
Montana	44	694,409	42	674,767
Nebraska	35	1,483,493	34	1,411,330
Nevada	48	488,738	50	285,278
New Hampshire	42	737,681	46	606,921
New Jersey	8	7,168,164	8	6,066,782
New Mexico	37	1,016,000	37	951,023
New York	2	18,236,967	1	16,782,304
North Carolina	12	5,082,059	12	4,556,155
North Dakota	46	617,761	45	632,446
Ohio	6	10,652,017	5	9,706,397
Oklahoma	27	2,559,229	27	2,328,284
Oregon	31	2,091,385	32	1,768,687
Pennsylvania	3	11,793,909	3	11,319,366
Rhode Island	39	946,725	39	859,488
South Carolina	26	2,590,516	26	2,382,594
South Dakota	45	665,507	41	680,514
Tennessee	17	3,923,687	17	3,567,089
Texas	4	11,196,730	6	9,579,677
Utah	36	1,059,273	38	890,627
Vermont	49	444,330	48	389,881
Virginia	14	4,648,494	14	3,966,949
Washington	22	3,409,169	23	2,853,214
West Virginia	34	1,744,237	30	1,860,421
Wisconsin	16	4,417,731	15	3,951,777
Wyoming	50	332,416	49	330,066

of Census, Final Report PC(1)–A1 United States

1950		1940		1930		1920	
Rank	Population	Rank	Population	Rank	Population	Rank	Population
17	3,061,743	17	2,832,961	15	2,646,248	18	2,348,174
51	128,643	51	72,524	51	[1]59,278	51	55,036
38	749,587	44	499,261	44	435,573	46	334,162
30	1,909,511	24	1,949,387	25	1,854,482	25	1,752,204
2	10,586,223	5	6,907,387	6	5,677,251	8	3,426,861
34	1,325,089	33	1,123,296	33	1,035,791	33	939,629
28	2,007,280	31	1,709,242	29	1,606,903	29	1,380,631
48	318,085	48	266,505	48	238,380	48	223,003
36	802,178	37	663,091	41	486,869	42	437,571
20	2,771,305	27	1,897,414	31	1,468,211	32	968,470
13	3,444,578	14	3,123,723	14	2,908,506	12	2,895,832
46	499,794	46	422,770	46	368,300	47	255,881
44	588,637	43	524,873	43	445,032	43	431,866
4	8,712,176	3	7,897,241	3	7,630,654	3	6,485,280
12	3,934,224	12	3,427,796	11	3,238,503	11	2,930,390
22	2,621,073	20	2,538,268	19	2,470,939	16	2,404,021
31	1,905,299	29	1,801,028	24	1,880,999	24	1,769,257
19	2,944,806	16	2,845,627	17	2,614,589	15	2,416,630
21	2,683,516	21	2,363,880	22	2,101,593	22	1,798,509
35	913,774	35	847,226	35	797,423	35	768,014
24	2,343,001	28	1,821,244	28	1,631,526	28	1,449,661
9	4,690,514	8	4,316,721	8	4,249,614	6	3,852,356
7	6,371,766	7	5,256,106	7	4,842,325	7	3,668,412
18	2,982,483	18	2,792,300	18	2,563,953	17	2,387,125
26	2,178,914	23	2,183,796	23	2,009,821	23	1,790,618
11	3,954,653	10	3,784,664	10	3,629,367	9	3,404,055
43	591,024	40	559,456	39	537,606	39	548,889
33	1,325,510	32	1,315,834	32	1,377,963	31	1,296,372
50	160,083	50	110,247	50	91,058	50	77,407
45	533,242	45	491,524	42	465,293	41	443,083
8	4,835,329	9	4,160,165	9	4,041,334	10	3,155,900
40	681,187	42	531,818	45	423,317	44	360,350
1	14,830,192	1	13,479,142	1	12,588,066	1	10,385,227
10	4,061,929	11	3,571,623	12	3,170,276	14	2,559,123
42	619,636	39	641,935	38	680,845	36	646,872
5	7,946,627	4	6,907,612	4	6,646,697	4	5,759,394
25	2,233,351	22	2,336,434	21	2,396,040	21	2,028,283
32	1,521,341	34	1,089,684	34	953,786	34	783,389
3	10,498,012	2	9,900,180	2	9,631,350	2	8,720,017
37	791,896	36	713,346	37	687,497	38	604,397
27	2,117,027	26	1,899,804	26	1,738,765	26	1,683,724
41	652,740	38	642,961	36	692,849	37	636,547
16	3,291,718	15	2,915,841	16	2,616,556	19	2,337,885
6	7,711,194	6	6,414,824	5	5,824,715	5	4,663,228
39	688,862	41	550,310	40	507,847	40	449,396
47	377,747	47	359,231	47	359,611	45	352,428
15	3,318,680	19	2,677,773	20	2,421,851	20	2,309,187
23	2,378,963	30	1,736,191	30	1,563,396	30	1,356,621
29	2,005,552	25	1,901,974	27	1,729,205	27	1,463,701
14	3,434,575	13	3,137,587	13	2,939,006	13	2,632,067
49	290,529	49	250,742	49	225,565	49	194,402

[1]Census taken as of October 1 of the preceding year.

vironmental limiting factors that prevent the reproductive potential from being realized.

FERTILITY: Pertaining to the birthrate of a population.

FERTILITY RATE: Number of live births per 1000 females 15 to 44 years of age.

INFANT DEATH RATE: Deaths per 1000 babies under the age of one year.

MORTALITY: The proportion of deaths to population.

NET REPRODUCTIVE RATE: Ratio between fertility and mortality.

POPULATION EXPLOSION: A pyramiding of a living population, especially the recent great increase in human numbers.

POPULATION PYRAMID: Graphical display of a population by age groups.

REPLACEMENT FERTILITY: Fertility resulting in a net reproductive rate of one.

REPRODUCTIVE POTENTIAL: The inherent property of an organism to reproduce, to survive, to increase in numbers; maximum reproductive power.

SPECTROPHOTOMETER: Instrument for measuring relative intensities of light in different parts of a spectrum.

TOTAL FERTILITY RATE: Average number of live children a woman will bear as she passes through her years of potential motherhood.

REFERENCES

Bogue, D. J. 1969. Principles of demography. John Wiley and Sons, Inc., New York.

Chandrasekhar, S. 1972. Infant mortality, population growth and family planning in India. University of North Carolina Press, Chapel Hill, North Carolina.

Clark, M. E. 1973. Contemporary biology: concepts and implications. W. B. Saunders Company, Philadelphia. pp. 508–513.

Commission on Population Growth and the American Future. 1972. A report on population growth and the American future. Superintendent of Documents, U.S. Government Printing Office, Washington, D.C.

Ebert, J. D., et al. 1973. Biology. Holt, Rinehart and Winston, Inc., New York. pp. 494–500.

Ehrlich, P. R., and A. H. Ehrlich. 1972. Population, resources, environment. W. H. Freeman and Company, San Francisco. 100 pp.

Frejka, T. 1973. The prospects for a stationary world population. Scientific American. *228*(3):15–23.

Harrison, G. A., and A. J. Boyce (eds.). 1972. The structure of human populations. Oxford University Press, Ely House, London.

Hauser, P. N. 1969. The population dilemma. Prentice-Hall, Inc., Englewood Cliffs, New Jersey.

Hertzbert, H. W. 1965. Teaching population dynamics. Population Instructional Materials Project, International Studies Program, Teacher's College, Columbia University, New York.

Kogan, B. A. 1970. Health: man in a changing environment. Harcourt Brace Jovanovich, Inc., New York. pp. 432–449.

Kormondy, E. J. 1969. Concepts of ecology. Prentice-Hall, Inc., Englewood Cliffs, New Jersey.

Meadows, D. H. 1972. Reckoning with recklessness. Ecology Today. January issue, page 47.

Population Reference Bureau. 1960. Population Bulletin. *16*(8):159. Population Reference Bureau, Inc., Washington, D.C.

Population Reference Bureau. 1967. Population profile. Population Reference Bureau, Inc., Washington, D.C.

Population Reference Bureau. 1970. Press release: annual world population growth reaches record high. Population Reference Bureau, Inc., Washington, D.C.

Population Reference Bureau. 1971. Man's population predicament. Population Bulletin. *27*(2):1–40. Population Reference Bureau, Inc., Washington, D.C.

Population Reference Bureau. 1972. The world population dilemma. Columbia Books, Inc., Washington, D.C.

Population Reference Bureau. 1973. Options: a study guide to population and the American future. Population Reference Bureau, Inc., Washington, D.C.

United Nations. 1952. Demographic yearbook. United Nations, New York.

United Nations. 1970. Total population estimates for world, regions and countries, each year, 1950–1985. Population Division Working Paper Number 34. United Nations, New York.

United States Bureau of the Census. 1970. Final report PC(1)–A1, United States summary. Superintendent of Documents, U.S. Government Printing Office, Washington, D.C.

United States Bureau of the Census. 1973. Rank of states by population: 1920–1970. Superintendent of Documents, U.S. Government Printing Office, Washington, D.C.

Wilson, E. O., et al. 1973. Life on earth. Sinauer Associates, Inc., Sunderland, Massachusetts. pp. 985–993.

study guide

drugs

minicourse

MIRHOB
VODKA
GOBBON'S
DISTILLED
LONDON DRY
GIN
THE HEART OF A GOOD COCKTAIL

Are the habits of our time very different from those of the past? Do we know more about the use, misuse, and abuse of drugs? Just what is a drug, and how do the so-called drugs do their thing?

PREREQUISITES: There are no prerequisites for this minicourse. However, it would be helpful if you are familiar with the functioning of the nervous system.

DRUG INFORMATION SURVEY

____ 1. A habit-forming drug differs from an addictive drug in that a habit-forming drug is one that produces
   a. psychological and physiological desire and a need to increase dosage.
   b. psychological and physiological desire but no need to increase dosage.
   c. psychological craving, which may lead to increased dosages but no physiological desire or need to increase dosage.
   d. steady use, but no desire or need to increase dosage.

____ 2. Drug addiction in the United States is most prevalent
   a. on high school and college campuses.
   b. in middle class suburban areas.
   c. in rural areas.
   d. in urban slum areas.

____ 3. Doctors prescribe barbiturates primarily
   a. to induce sleep.
   b. to reduce tension.
   c. to relieve pain.
   d. in research.

____ 4. A baby born to a mother addicted to opiates will
   a. also be addicted.
   b. have a high probability of physical deformity.
   c. have a high probability of mental retardation.
   d. be unaffected.

____ 5. Which of the following is not one of the major conclusions of the Surgeon General's report on smoking?
   a. The habitual use of tobacco is related primarily to physiological drives, reinforced and perpetuated by promotional efforts of the tobacco industry.
   b. Cigarette smoking is associated with a 70 percent increase in the age-specific death rate of men (data for women similar).
   c. The total number of excess deaths causally related to cigarette smoking in the U.S. population cannot be accurately estimated.
   d. Cigarette smoking contributes substantially to deaths from certain specific diseases and to the overall death rate.

____ 6. Which of the following sets of terms refer to amphetamines?
   a. "Red devils," Amytal, Seconal
   b. Benzedrine, "Speed," "Dexies"
   c. Librium, Thorazine, Compazine
   d. Demerol, Methadone, Codeine

____ 7. Amphetamines may enable a person to
   a. think more clearly.
   b. do better on exams.
   c. stay awake.
   d. remain calm in a pressure situation.

_____ 8. The greatest health danger from amphetamine use is its effect on
a. body temperature.
b. the respiratory center.
c. heartbeat.
d. O_2–CO_2 balance in the blood.

_____ 9. The most important medical use of hallucinogens is in
a. treating depression.
b. enhancing creativity and perception.
c. overcoming fear and anxiety.
d. research on human behavior.

_____ 10. How does LSD affect vision and hearing?
a. It affects how sights and sounds are experienced.
b. It reduces the sensitivity of sight and hearing.
c. It makes the eyes see better and the ears hear better.
d. It has no effect on vision and hearing.

11. The most widely used drug in the U.S. is _________.

12. The most widely abused drug in the U.S. is ________.

_____ 13. Which is the best statement concerning possible physiological damage from marijuana?
a. It is physiologically harmful only if used often.
b. It has been proved to be less harmful physiologically than other commonly used drugs.
c. There is evidence to show that regular smoking of marijuana weakens the body's immune defenses against disease.
d. Its use does no physical damage to the body.

_____ 14. Which of the following best accounts for the belief that use of marijuana causes use of more dangerous drugs?
a. Frequent use of marijuana causes a need for addictive drugs.
b. A marijuana user is likely to contact and be influenced by users of other drugs.
c. Frequent use of marijuana increases desire for other behavior-modifying drugs.
d. Marijuana weakens willpower and triggers experimentation with other drugs.

_____ 15. An individual who occasionally uses more than the prescribed amount of addictive drugs
a. will not become addicted.
b. may become addicted.
c. will become addicted in time.
d. is already moderately addicted.

_____ 16. Which of the following is the best description of a likely candidate for drug addiction?
a. an individual unable to achieve a satisfactory social adjustment
b. an individual with tendencies toward antisocial behavior
c. no one kind of person
d. an individual lacking self-discipline

_____ 17. Death of an addicted person from drug abuse is most likely to result following discontinuance of which of the following?
a. alcohol
b. heroin
c. secobarbital
d. amphetamine

_____ 18. Black market drugs are dangerous primarily because
a. the dosage is often unknown, as is the purity.
b. they tend to be stronger.
c. they have a greater tendency to cause addiction than prescribed drugs.
d. they are often spoiled drugs that have been discarded.

I. WHAT IS A DRUG?

A. Dictionary Definitions of the Term Drug

Webster's Second New International Dictionary, 1950

DRUG: Any substance used as a medicine, or in making medicines, for internal or external use. According to the Pure Food and Drug Act the term drug includes all medicines and preparations recognized in the *United States Pharmacopoeia and National Formulary* for internal or external use, and any substance or mixture of substances intended to be used for the cure, mitigation, or prevention of disease of either man or other animals.

(By permission. From *Webster's Second New International Dictionary* © 1950 by G. & C. Merriam Company, Publishers of the Merriam-Webster Dictionaries.)

Webster's New Collegiate Dictionary, 1975

DRUG: A substance used in dyeing or chemical operations; a substance intended for use in the diagnosis, cure, or prevention of disease; a substance that causes addiction or habituation.

(By permission. From *Webster's New Collegiate Dictionary* © 1975 by G. & C. Merriam Company, Publishers of the Merriam-Webster Dictionaries.)

B. In This Minicourse, The Following Terms Are Used as Defined Here:

DRUG—Any biologically active substance that is deliberately introduced into the body to affect its functioning.

USE—Taking a substance for the purpose for which it was intended, in the prescribed frequency and dosage.

MISUSE—Taking a substance for the purpose for which it was intended, but *not* in the prescribed frequency and dosage.

ABUSE—Deliberate taking of a substance for other than its intended purpose, and in a manner that can result in damage to health or ability to function.

II. WHAT ARE THE PATTERNS OF DRUG USE AND ABUSE?

Figure 1. Typical users and abusers??

Figure 2. Are you a drug user?

A. Possible Consequences of Drug Use and Abuse

1. *Habituation:* Psychological dependence or compulsion to continue use of a drug. Deprivation causes restlessness and anxiety but not withdrawal illness.

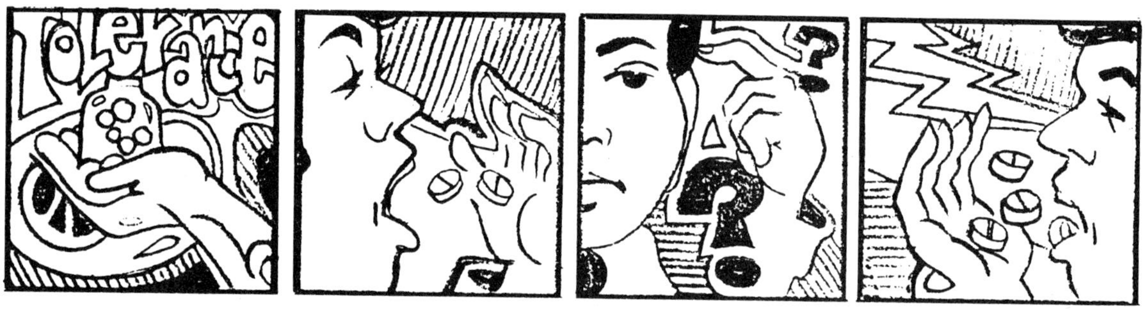

2. *Tolerance:* Bodily adaptation over a period of time requires larger and larger doses to produce the same effect a smaller dose initially produced.

3. *Addiction:* The body requires continual administration of a drug in order to avoid physical pain of withdrawal. Tolerance usually accompanies addiction.

Figure 3. Some consequences of drug use and abuse. (Adapted from CRM Books, *Life and Health,* Copyright © 1972 by Ziff-Davis Publishing Company.)

B. Major Factors Influencing Use and Abuse of Drugs

1. PSYCHOLOGICAL—personality and character structure of the drug taker
   a. desire for pleasure, euphoria, relaxation, socialization
   b. desire to escape from reality, from boredom, to relieve pain, escape anxiety
   c. desire to expand creative potential
   d. emotional immaturity or inadequacy
   e. lack of self-discipline
2. SOCIOLOGICAL—life style of the drug taker
   a. excessive prescription and promotion by doctors and the drug industry
   b. emulating one's parents and peers
   c. rebellion against or escape from society
   d. to gain acceptance to a group
3. PHARMACOLOGICAL—the chemical effect of the drug on the drug taker
   a. drug's capacity to produce tolerance and addiction
   b. fear of withdrawal symptoms

The effect of a drug is determined to a great extent by the nature of the person who takes the drug and the nature of the situation in which it is being taken.

Drug experts agree there is no single reason why a person uses drugs, no single pattern of abuse.

The Deadly Downer

According to its manufacturers, methaqualone is a dependable and effective sleeping pill. Blurbs in the standard *Physicians' Desk Reference* attest to its "sedative and hypnotic" effects and its ability to induce prompt sleep, but warn that the drug may also produce dependency. The warning is appropriate. Though the drug may be safe if it is taken as prescribed by a physician, increasing numbers of Americans—especially on campuses and in ghettos—are obtaining the drug illegally and taking it indiscriminately. Methaqualone is rapidly becoming one of the most popular —and dangerous—drugs of abuse in the U.S. The Government has only recently recognized its dangers and moved toward establishing controls.

On the street and on campus, methaqualone is known by various corruptions of its trade names: "quads" (from Quaalude, made by William H. Rorer Inc.), "French Quaalude" (from mispronunciation of Parest, the Parke, Davis & Co. brand name), and as "soapers" (from Sopor, made by Arnar-Stone Laboratories Inc.). It was so popular among the young people who camped out at last year's political conventions that Miami Beach's Flamingo Park was dubbed "Quaalude Alley." Its abuse is now a major problem at many colleges and high schools. It is currently favored by the addict community in the New York City borough of The Bronx, and is so fashionable among some drug culturists that bowls of it have replaced peanuts as a cocktail-party staple.

Users—who are mostly youngsters —are rhapsodic about the euphoric, spaced-out state the drug can produce. Many, calling methaqualone "heroin for lovers," also believe that it is an aphrodisiac. They are mistaken. As a "downer," or depressant, the drug may release the user from his normal sexual restraints. But it is also likely to make a male incapable of normal sexual performance. As one Vassar man puts it, "All your inhibitions are definitely broken down—like everything else in your body."

The drug's effects on the body can, in fact, be deadly. Most doctors prescribe methaqualone in capsules of 150 mg. to 300 mg., one of which should be enough to put the average adult to sleep. But drug abusers are not content with such tame dosages. Their usual pattern is to take two or more, and then wash them down with a few pints of beer.

The combination can be catastrophic. Methaqualone and alcohol are synergistic: one multiplies the effects of the other. Taken together, they can depress the respiratory center and stop breathing; they may also slow the reflexes in the back of the throat, so that if a user vomits (which may happen after a large dose of methaqualone), he can choke and die.

No Controls. On top of this, methaqualone is addictive. Unlike insomniacs who abide by their doctor's prescriptions, abusers develop a tolerance for the drug and begin taking increasingly larger doses. Then they may become addicted. "Resistance can develop after only four days," says Mike, a 20-year-old San Franciscan. "Then it takes four to do what one used to do for you. The withdrawals are much worse than heroin, with the same kind of convulsion as in an epileptic fit."

Despite these dangers, however, controls over methaqualone are virtually nonexistent. Federal authorities have been considering regulations for methaqualone for the past few months, but have not yet proposed any controls on production or sales. The individual states have also been slow in acting. In New York, methaqualone is classed as a dangerous drug; that means that pharmacists cannot refill the prescription without written authorization from a doctor. Connecticut and Maryland are considering controls. But in the rest of the country, the drug and its dangers are unrestricted.

BY STEWART ALSOP

I DIDN'T STOP IN BALTIMORE

WASHINGTON—I woke up suddenly, feeling very wide awake, pulled the light cord, and sat straight up in bed.

"We'll be stopping in Baltimore," I announced loudly, in an authoritative voice. I looked around me. A private car, evidently. Good of the railroad people to supply those fruits, and all those flowers. The furniture looked a bit shoddy, but what can one expect these days? My wife was no doubt in the next stateroom. Where was the connecting door? I got up to explore.

The car was swaying heavily—the roadbed must be a disgrace. I supported myself on a table, and then a desk. Then there was a space of empty floor. I was halfway across it when the car gave a lurch, and I fell down. I sat on the floor for a bit, getting my bearings, then scrambled to my feet again, and opened a narrow green door. A locker, with my own clothes in it. I opened another door—a small bathroom.

Then I came to a much bigger door, and opened it, and leaned against the doorjamb. The swaying had stopped —the train, apparently, had halted. Outside was what I assumed was the Baltimore station—wide platform, dim lights, green tile. A whimpering noise, then silence, and no one to be seen. There was something hellishly grim about the place. Suddenly I was quite sure I didn't want to stop in Baltimore.

"We won't stop here," I said, again in a firm, authoritative voice. "Start up the train, and carry on."

SAFE

I turned back toward my bed, and the big door closed behind me. I fell down twice on the way back—the crew must be pouring on the power, I thought—and getting into bed was like mounting a bucking horse. Safe in bed, I turned off the light, and was asleep in an instant.

This small episode did not happen on a private railroad car. (Who the hell did I think I was? Some ancient Belmont or Vanderbilt?) It happened in my room on Floor 12, the Solid-Tumor Ward of the cancer clinic of the National Institutes of Health. My room, a room I'd been in for going on eight weeks, was the private car, and the corridor of the Solid-Tumor Ward was the Baltimore platform. But the falling down was real enough—the next morning I had the bruises to prove it.

I was in the Solid-Tumor Ward because NIH, for mysterious bureaucratic reasons, has ended the adult leukemia program, and the adult leukemics who are still about are taken care of in other wards. Leukemia is a sneaky disease—it attacks by indirection. It weakens the blood cells that fight infection, and then sends in surrogate diseases to finish off the victim.

In the two months I was in the Solid-Tumor Ward, my leukemia (or whatever it is—it is a most atypical disease) tried hard to finish me off. I had pneumonia; then an infection of the lower intestines; then a lung clot; then an edema of the lungs; then a second pneumonia; then two minor operations and a major operation; then another infection.

STUBBORN

The major operation—opening my chest and snipping off a thumbnail-size bit of lung—was necessary because pneumonia can't be treated unless the doctors know what kind it is, and my second pneumonia stubbornly refused to identify itself. The bit of lung was given every known test—and the brilliant doctors at NIH know every known test—but it still refused to identify itself. Meanwhile the lethal "infiltrate" continued relentlessly to spread across both lungs, and my wife was told that the prognosis was "grim."

It was four days after the operation that I decided not to stop in Baltimore. The next day the able young doctor in charge of my case, Jack Macdonald, told me that he might be imagining things, but the X-rays of my lungs looked a bit better—certainly no worse. The day after, he said there was no doubt about it—the infiltrate was receding. Some days later, my battered old lungs were as close to normal as they will ever be.

Why? Jack Macdonald and the other doctors say frankly they don't know, though they all have a favorite guess. I have a favorite guess, too. My guess is that my decision not to stop at Baltimore had something to do with it. In a kind of fuzzy, hallucinated way, I knew when I announced the decision that it was a decision not to die.

Death is a word rarely mentioned in the NIH cancer clinic. But one is more aware of its presence in the Solid-Tumor Ward than in the now defunct leukemia ward. Leukemia is a sneakier disease than solid-tumor cancer, but it is also kinder at the end. Most leukemics drift into a quiet death, while the victims of a solid tumor can (although they by no means always do) suffer agonies before death comes.

All the rooms at NIH are double rooms, and almost all the patients (except those thought soon to die, which may be why I was alone on my private car) have roommates. Several of my roommates were terminal solid-tumor patients, who were given opium-based analgesics or painkillers at predetermined intervals, usually four hours. Almost always, the analgesic wore off too soon. To hear grown men, and brave men, whimpering, or howling, or pleading with a helpless nurse for another "shot" is not an experience likely soon to be forgotten.

As a result of that experience, I reached certain conclusions. NIH, like most hospitals, goes to great lengths to prevent suicide—patients are forbidden pills of their own, and the windows and even the rooftop solarium are firmly screened. It seems to me that a patient suffering beyond endurance should be given the option of ending his own life, and the means to do so should be supplied on request. An unquestionably terminal patient should be given another option. He should be given as much painkilling drug as he himself feels he needs. The drug probably ought to be heroin, which is estimated to be about four times as effective an analgesic as the synthetic painkillers now mostly in use. If a human being must die, it is surely better that he die in the illusion of painless pleasure—and heroin is very pleasurable—than in lonely agony.

MYSTERIES

Another conclusion is harder to define. Shakespeare, as usual, came closest, in the familiar cliché-quote from Hamlet: "There are more things in heaven and earth, Horatio, than are dreamt of in your philosophy." Perhaps my decision not to stop in Baltimore had nothing to do with my astonishing recovery. But there are mysteries, above all the mystery of the relationship of mind and body, that will never be explained, not by the most brilliant doctors, the wisest of scientists or philosophers.

For the rest, I hope to put back some (but not all) of the 43 pounds I have lost, to get a bit of rest, to get rid of the tail end of that second infection, and then again to examine in this space the more mundane mysteries of political Washington.

Here is a portion of the published report of a biology professor, describing his experiences following ingestion of a nightshade plant preparation. The author was taken to a hospital, placed in a barred cell, and tied to a bed.

Most rooms usually have rectangular shuttered air vents situated either near the ceiling or floor. The vent in my cell was located just above the door. To my amazement it began to fill up like a small football stadium with tiered rows of my former students. They were wearing brightly colored berets and horn rimmed glasses. Each had a bottle of coke and a bag of popcorn. They single filed into the bleachers, finally filling it to capacity. Now they sat very rigid without appearing to notice my presence. However, when I turned my head as if to look at some other area of the room the scene suddenly became highly animated. As they guzzled their cokes and stuffed popcorn into their mouths they pointed their fingers at me and laughed sardonically. When ever I focused my eyes on their antics, they suddenly froze and took no notice of me. It is of interest to note that I recognized every face in the group and that I had given each and every one of them the D or F grade in beginning Biology.

C. E. JOHNSON. 1967.

Figure 4. An epitaph.

III. HOW ARE DRUGS CLASSIFIED?

TABLE 1.

Classification of drugs.*

NAME OF DRUG OR CHEMICAL	SLANG NAMES	USUAL SINGLE ADULT DOSE	DURATION OF ACTION (HOURS)	METHOD OF TAKING	CURRENT MEDICAL USES
I. DEPRESSANTS, SEDATIVES	Downers		4	Swallowing pills or capsules	Treatment of insomnia and tension; induction of anesthesia
Barbiturates	Barbs	50–100 mg			
Amytal	Blue devils				
Nembutal	Yellow jackets, dolls				
Seconal	Red devils				
Phenobarbital	Phennies				
Doriden (glutethimide)	Goofers	500 mg			
Chloral hydrate		500 mg			
Miltown, Equanil (meprobamate)		400 mg			
Methaqualone	Soapers, quads				
Alcohol	Booze, hooch, juice	1-1/2 oz. gin or whiskey, 12 oz. beer	2–4	Swallowing liquid	Rare; sometimes used as a sedative (for tension)
Whiskey					
Gin					
Beer					
Wine					
II. NARCOTICS (OPIATES, ANALGESICS)					Treatment of severe pain, diarrhea, and cough
Opium	Op	10–12 "pipes" (Asia)	4	Smoking (inhalation)	
Heroin	Horse, H, smack, junk	Variable—bag or paper with 5–10% heroin		Injecting in muscle or vein	
Morphine		15 mg			
Codeine		30 mg			
Percodan		1 tablet			
Demerol		50–100 mg			
Methadone	Dolly	2–4 oz. (for euphoria)		Swallowing	
Cough syrup (*Cheracol, Hycodan,* etc.)					
III. STIMULANTS	Uppers				Treatment of obesity, narcolepsy, fatigue, depression
Amphetamines	Pep pills	2.5–5.0 mg	4	Swallowing pills or capsules or injecting in veins	
Benzedrine	Bennies, cartwheels				
Methedrine	Crystal, speed, meth				
Dexedrine	Dexies				
Cocaine	Coke, snow	Variable		Sniffing or injecting	Anesthesia of the eye and throat
Preludin					
Caffeine	Java		2–4		Mild stimulant; treatment of some kinds of coma
Coffee, tea, *Coca-Cola*		1–2 cups or 1 bottle		Swallowing liquid	
No-Doz, APC		5 mg		Swallowing pills	
Nicotine (and coal tar)	Fag	1 cigarette	1–2	Smoking (inhalation)	None (used as an insecticide)
Cigarettes, cigars					
IV. HALLUCINOGENS					Experimental study of mind and brain function; enhancement of creativity and problem solving; treatment of alcoholism, mental illness, and the dying person
LSD	Acid, sugar	150 micrograms	10–12	Swallowing liquid, capsule, pill (or sugar cube)	
Psilocybin	Mushrooms	25 mg	6–8		
STP		5 mg			
DMT		5 mg		Smoking	
Mescaline (peyote)	Cactus	350 mg	12–14	Chewing plant	
V. CANNABIS	Pot, grass, tea, weed, stuff, hash, joint, reefer	Variable—1 cigarette or pipe or 1 drink or cake (India)	4		None
Marijuana				Smoking (inhalation)	
Hashish				Swallowing	
VI. TRANQUILIZERS			4–6	Swallowing pills or capsules	Treatment of anxiety, tension, alcoholism, neurosis, psychosis, psychosomatic disorders, and vomiting
Librium (chlordiazepoxide)		5–10 mg			
Phenothiazines					
Thorazine		10–25 mg			
Compazine		10 mg			
Stelazine		2 mg			
Reserpine (rauwolfia)		1 mg			
VII. ANTIDEPRESSANTS			4–6	Swallowing pills or capsules	Treatment of moderate to severe depression
Ritalin		10 mg			
Dibenzapines (*Tofranil, Elavil*)		25 mg, 10 mg			
MAO inhibitors (*Nardil, Parnate*)		15 mg, 10 mg			
VIII. MISCELLANEOUS					None except for antihistamines used for allergy, amyl nitrite for some episodes of fainting, and nitrous oxide for anesthesia
Glue, gasoline, and solvents		Variable	2	Inhalation	
Amyl nitrite		1–2 ampules			
Antihistamines		25–50 mg			
Nutmeg		Variable		Swallowing	
Nonprescription "sedatives" (*Compoz*)					
Nitrous oxide					

Italics refer to registered tradenames.

*Adapted from *The Pleasure Seekers,* Copyright © 1969, by Joel Fort, reprinted by permission of the publisher, The Bobbs-Merrill Company, Inc.

TABLE 1.

Classification of drugs. *(Continued)*

	POTENTIAL FOR PSYCHOLOGICAL DEPENDENCE (HABITUATION)	TOLERANCE POTENTIAL	POTENTIAL FOR PHYSICAL DEPENDENCE (ADDICTION)	OVERALL ABUSE AND TOXICITY POTENTIAL*	USUAL SHORT-TERM EFFECTS (PSYCHOLOGICAL, PHARMACOLOGICAL, SOCIAL)**	USUAL LONG-TERM EFFECTS (PSYCHOLOGICAL, PHARMACOLOGICAL, SOCIAL)
I.	High	Yes	Yes	High	CNS depressants; sleep induction; relaxation (sedation); sometimes euphoria; drowsiness; impaired judgment, reaction time, coordination, and emotional control; relief of anxiety-tension; muscle relaxation	Irritability, weight loss, addiction with severe withdrawal illness (like DTs); habituation, addiction
	High	Yes	Yes	High	CNS depressant; relaxation (sedation); sometimes euphoria; drowsiness; impaired judgment, reaction time, coordination, and emotional control; frequent aggressive behavior	Diversion of energy and money from more creative and productive pursuits; habituation; possible obesity with chronic excessive use; irreversible damage to brain and liver; addiction with severe withdrawal illness (DTs), possible death
II.	High	Yes	Yes	High	CNS depressants; sedation, euphoria, relief of pain, impaired intellectual functioning and coordination	Constipation, loss of appetite and weight, temporary impotency or sterility; habituation, addiction with unpleasant and painful withdrawal illness
III.	High	Yes	No	High	CNS stimulant; increased alertness; reduction of fatigue, loss of appetite, insomnia, often euphoria	Restlessness, irritability, weight loss, toxic psychosis (mainly paranoid); habituation
	Moderate	Yes	No	Minimal	CNS stimulant; increased alertness, reduction of fatigue	Sometimes insomnia, restlessness or gastric irritation; habituation
	High	Yes	No	High	CNS stimulant; relaxation (or distraction)	Lung (and other) cancer, heart and blood vessel disease, cough, etc.; habituation; diversion of energy and money; air pollution; fire
IV.	Minimal	Yes (rare)	No	Moderate	Production of visual imagery, increased sensory awareness, anxiety, nausea, impaired coordination; sometimes consciousness-expansion	Usually none; sometimes precipitates or intensifies an already existing psychosis; more commonly can produce a panic reaction
V.	Moderate	No	No	Minimal to Moderate	Relaxation, euphoria, increased appetite, some alteration of time perception, possible impairment of judgment and coordination	Usually none; habituation; occasional acute panic reactions
VI.	Minimal	No	No	Minimal	Selective CNS depressants; relaxation, relief of anxiety-tension; suppression of hallucinations or delusions; improved functioning	Sometimes drowsiness, dryness of mouth, blurring of vision, skin rash, tremor, occasionally jaundice, agranulocytosis, or death
VII.	Minimal	No	No	Minimal	Relief of depression (elevation of mood), stimulation	Basically the same as Tranquilizers, above
VIII.	Minimal to Moderate	Not Known	No	Moderate to High	When used for mind alteration, generally produces a "high" (euphoria) with impaired coordination and judgment	Variable—some of the substances can seriously damage the liver or kidney and some produce hallucinations

*Excessive, often compulsive use of a drug to an extent that it damages an individual's health or social or vocational adjustment, or is otherwise specifically harmful to society.

**Always to be considered in evaluating the effects of these drugs is the amount consumed, frequency, time interval since ingestion, food in the stomach, combinations with other drugs, and, most important, the personality or character of the individual taking it and the setting or context in which it is taken. The determinations made in this chart are based upon evidence with human use of these drugs rather than upon isolated, artificial-experimental situations or animal research.

IV. HOW DO DRUGS AFFECT THE NERVOUS SYSTEM?

A. How Does the Nervous System Coordinate Movement?

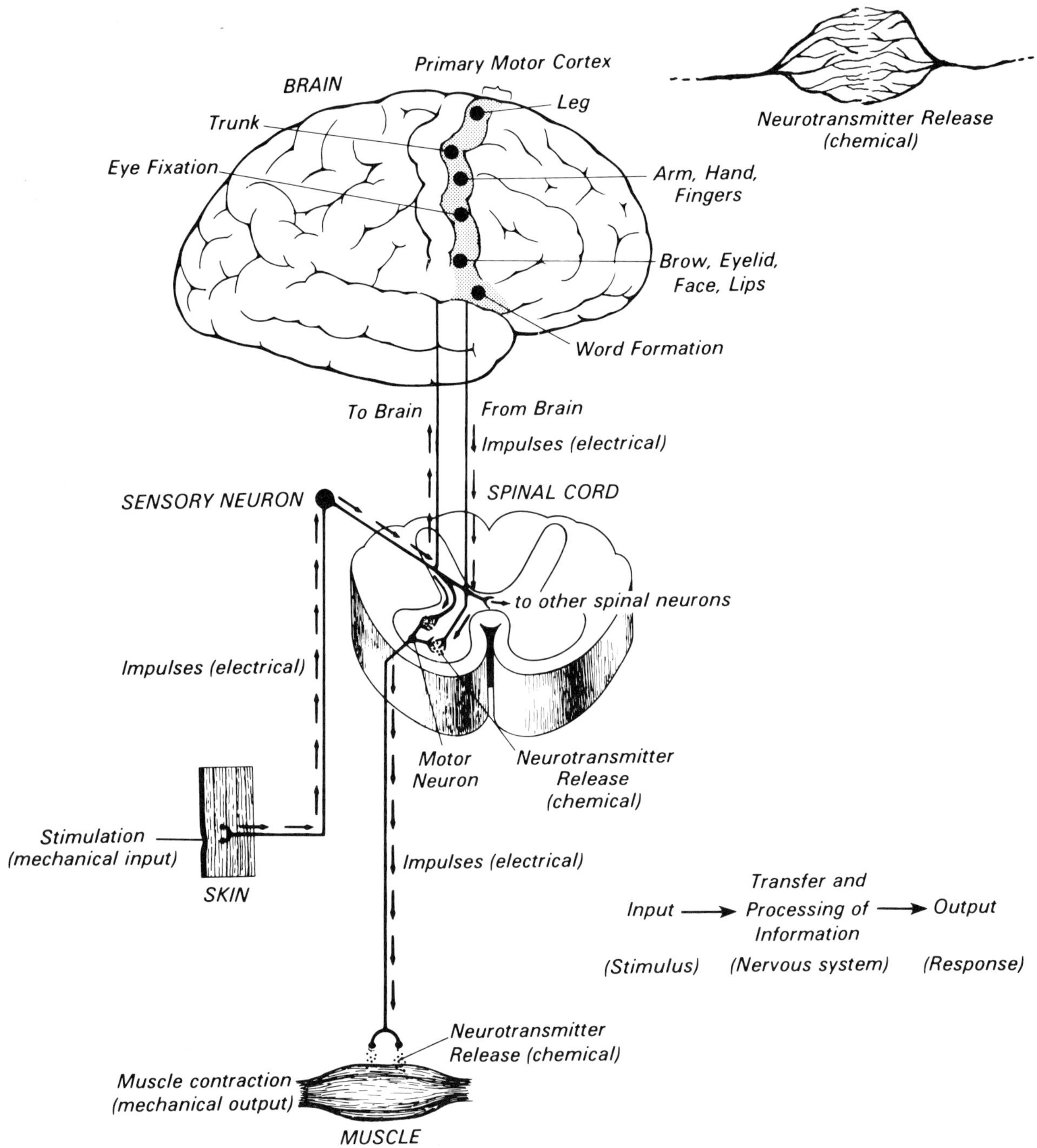

Figure 5. Summary diagram illustrating the transfer of sensory information and its processing in the spinal cord and brain, leading to an output of muscle contraction, the response.

EXERCISE 1. What are some simple nerve pathways?

Use the summary diagram in Figure 5 to answer the following questions.

(1) What pathway would an impulse follow if you were jabbed in the arm with a sharp instrument? (Your response as you jerk your arm away would be a *reflex* action. Describe the pathway from input to output.)

(2) What pathway would an impulse follow if you were touching a warm object and you decided to move your arm away from the object? Describe the pathway.

(3) What would happen along the pathway if the release of a chemical neurotransmitter between successive neurons was being inhibited by something?

B. What Are the Major Subdivisions and Functions of the Human Brain?

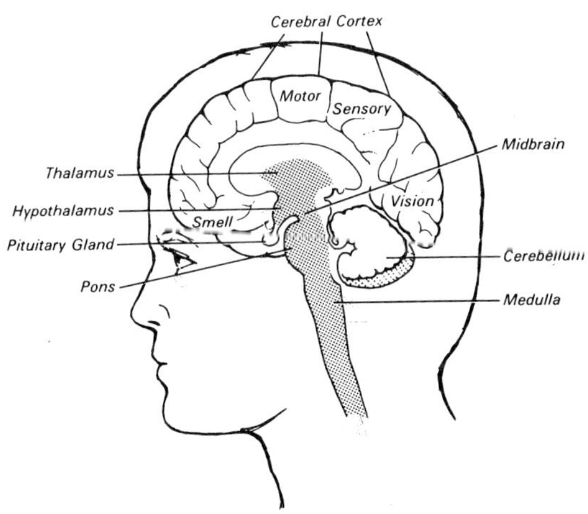

Figure 6. Middle section of the human brain. (Hearing center on the outside has been cut away.) The components of the *brainstem* are shaded more darkly. (After Clark. 1973.)

1. medulla—controls unconscious functions such as heart rate, blood pressure, and breathing reflexes.
2. cerebellum—controls muscle coordination and maintains posture, muscle tone, and balance.
3. pons—contains nerve pathways joining the medulla and higher regions of the brain (cerebral cortex).
4. midbrain—serves as a relay for ascending and descending nerve tracks, which branch and connect with other nerve fibers along the way.
5. thalamus—a relay center that serves to integrate sensory impulses; also serves in the integration of emotional behavior.
6. hypothalamus—concerned with the regulation of endocrine gland function and autonomic nerve activity; basal control for bodily functions and metabolism. (Controls basic drives such as hunger, thirst, and sex; regulates release of growth and other metabolic hormones.)
7. cerebral cortex—seat of higher mental activity; abstract thinking, voluntary movement, memory; location for specific sensory and motor regions. Basal ganglia are located in the inner part of the cerebrum.

C. How Do Drugs Interfere with Normal Brain Functions?

1. functioning of the basal ganglia

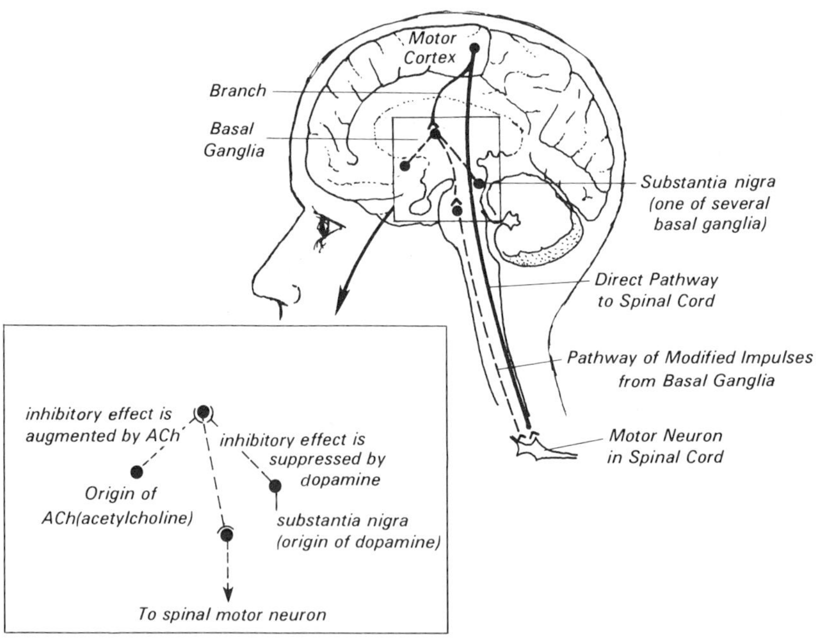

Figure 7. Cortical motor control and the basal ganglia. Two sets of motor impulses are sent to spinal motor neurons. One pathway directly connects the motor cortex to the spinal neuron. A second pathway, from the basal ganglia, also affects spinal motor neurons. Impulses from the basal ganglia have an inhibiting function, which is augmented by acetylcholine (ACh) and suppressed by dopamine. (After Clark. 1973.)

EXERCISE 2. How does the basal ganglia modify nerve impulses?

Nerve impulses from the basal ganglia *inhibit* spinal motor neuron activity. The inhibitory effect of the basal ganglia is *increased* by a release of acetylcholine (ACh) and is *decreased* by a release of dopamine from the substantia nigra. Normal coordinated muscle activity is maintained by *balanced* secretions of acetylcholine and dopamine.

What would be the effect on body function (muscle coordination) if the substantia nigra is unable to manufacture dopamine?

TYROSINE —Enzyme 1→ L-DOPA —Enzyme 2→ DOPAMINE

↑ (Enzyme 1) Defective in Parkinson's Disease

Figure 8. Steps in dopamine production.

2. functioning of the reticular formation

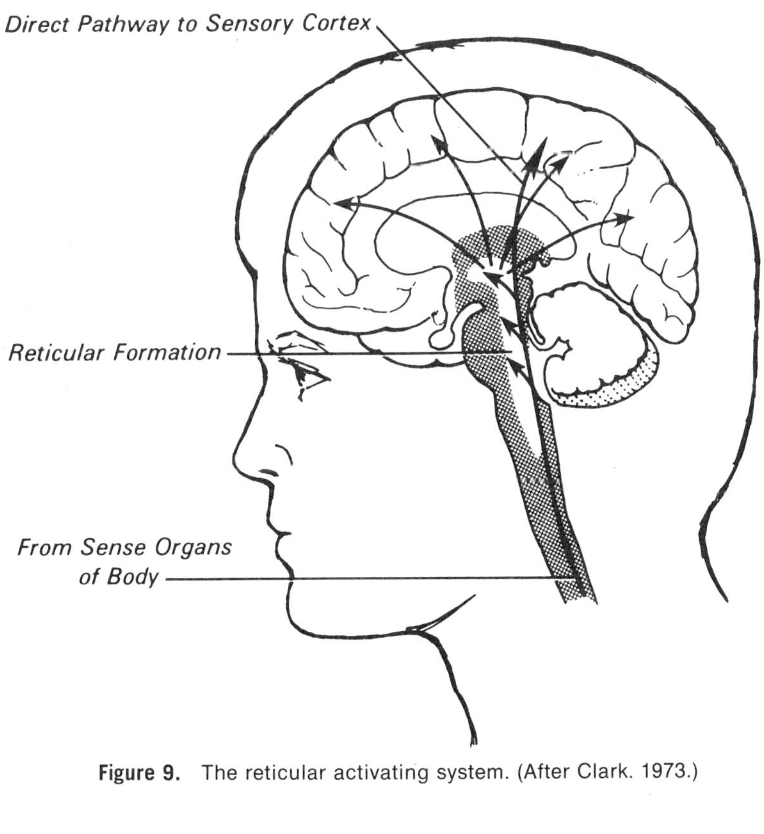

Figure 9. The reticular activating system. (After Clark. 1973.)

EXERCISE 3. How does the reticular activating system function?

Sensory nerves in the reticular formation branch before passing on to the cortex. The function of this reticular system is to sort out the significant impulses while filtering "background noise."

Three neurotransmitters appear to be involved in the system function:

a. 5-HT (5-Hydroxytryptamine) increases drowsiness and sleep

b. Noradrenalin increases awake condition

c. Acetylcholine diminishes resting potential of brain cortex cells, allowing more frequent impulses during periods of stimulation

What features would you look for in designing a drug that would stimulate wakefulness? Cause sleep?

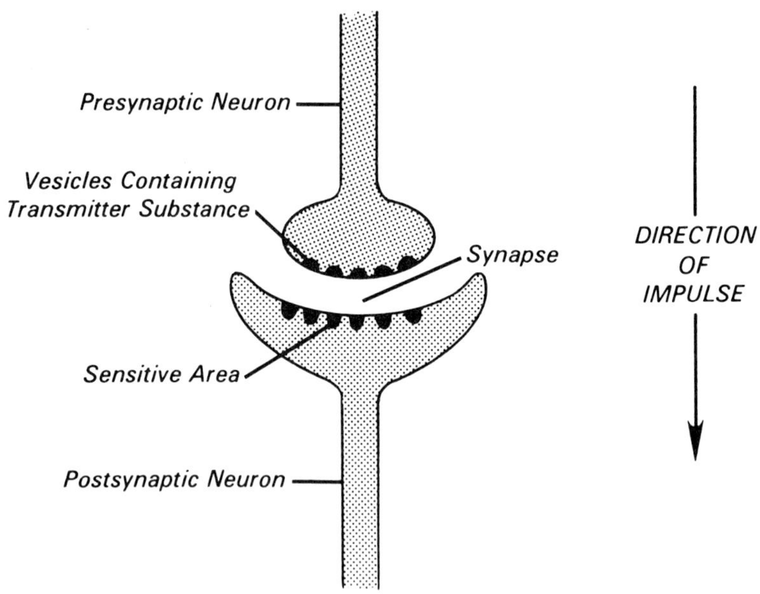

Figure 10. Synaptic region of a nerve pathway.

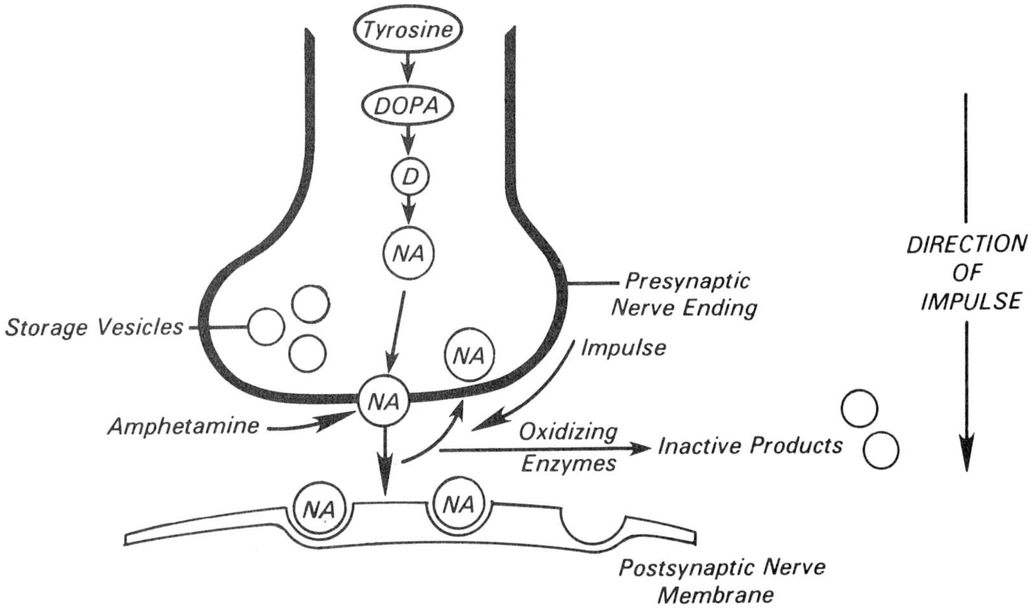

Figure 11. Effect of amphetamine at nerve synapse. Nerve impulse causes noradrenalin (NA) to be released at the presynaptic nerve ending. This release is apparently enhanced by amphetamines, increasing the chance that NA will reach the postsynaptic nerve ending continuing the nerve impulse. (After Rech, R. H., and Moore, K. E., 1971, *An Introduction to Psychopharmacology.* Raven Press. New York.)

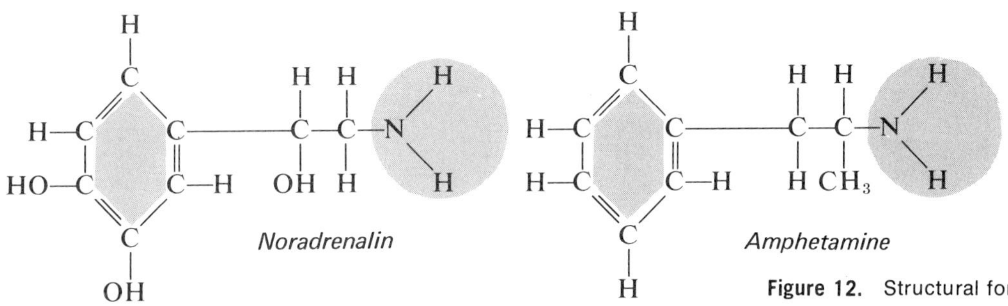

Figure 12. Structural formulas.

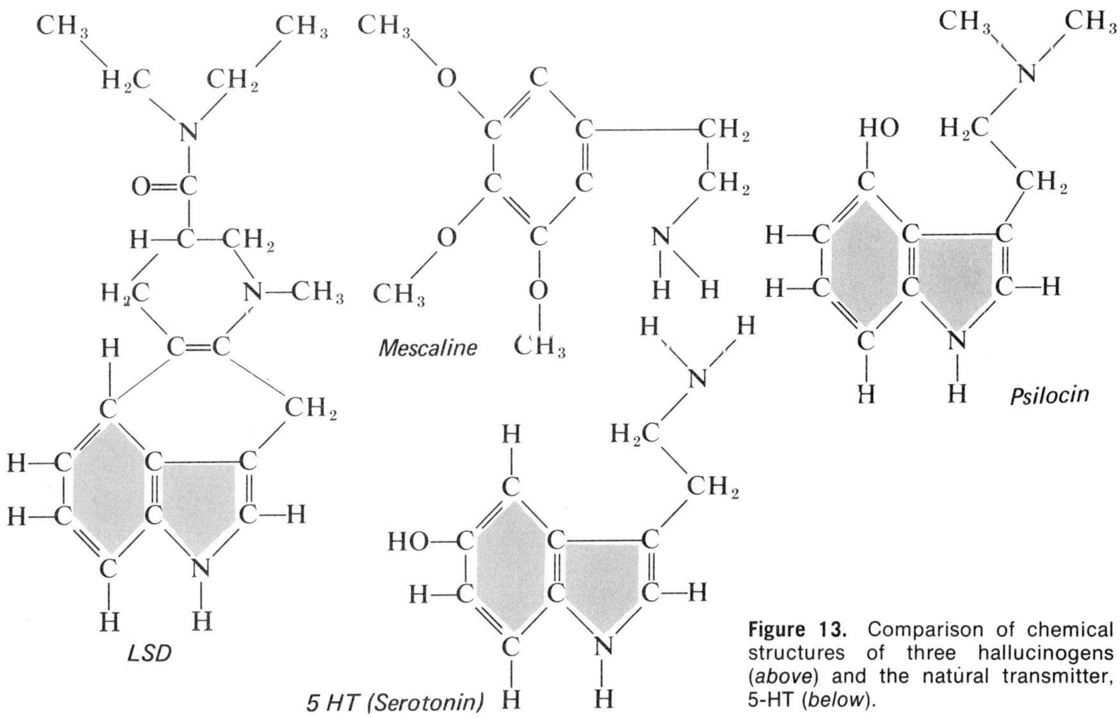

Figure 13. Comparison of chemical structures of three hallucinogens (*above*) and the natural transmitter, 5-HT (*below*).

V. WHAT ARE THE EFFECTS OF SOME SPECIFIC DRUGS AND HOW ARE THESE EFFECTS PRODUCED?

A. Alcohol

Figure 14. "Isn't it disgraceful how these kids nowadays are always getting high on drugs? They just don't realize what they're doing to their bodies."

1. the pathway of ingested alcohol

Youths are moving from a wide range of other drugs to the most devastating drug—the one most widely misused of all—alcohol.

Dr. Morris Chafetz
Director, Dept. HEW
National Institute on Alcohol Abuse and Alcoholism (NIAAA)

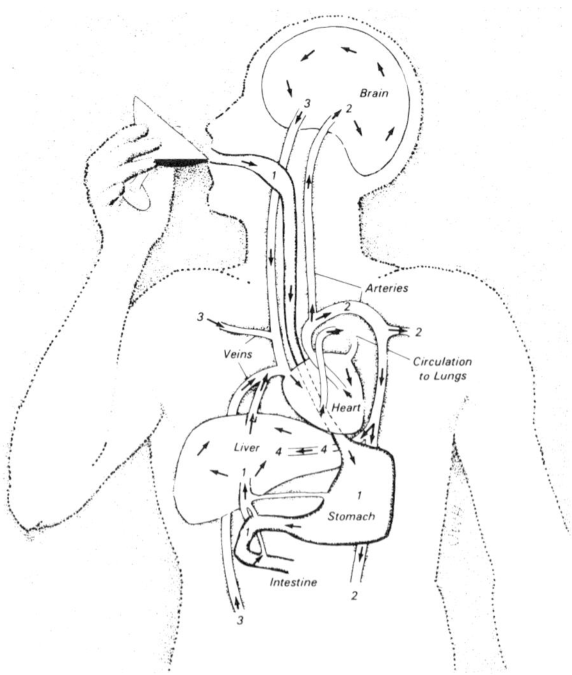

Figure 15. The distribution of alcohol in the body. (From Clark. 1973.)

a. Alcohol is absorbed directly into the bloodstream from the stomach and intestine.

b. The blood flows to the liver, which begins the metabolic breakdown of alcohol. The excess alcohol recirculates to various body parts.

c. The alcohol returns to the heart and is recirculated.

d. The liver removes and metabolizes some of the alcohol at each circulation.

ALCOHOL $\longrightarrow$ ACETALDEHYDE $\longrightarrow$ ACETATE $\longrightarrow$ CO_2 + WATER

Figure 16. The breakdown of alcohol by the liver.

2. What are some behavioral effects of alcohol?

   a. factors affecting blood alcohol levels:
      (1) consumption—amount of alcohol ingested
      (2) absorption—rate at which alcohol enters the blood
      (3) dilution—volume of total body water
      (4) metabolism—rate of alcohol breakdown in the liver
      (5) elimination—rate of excretion (in lungs and urine) plus metabolism

   b. some effects on the brain and behavior

Figure 17. The effect of blood alcohol levels on behavior.

B. Cigarettes

1. What are some physiological effects of smoking?

EXERCISE 4. What is the effect of smoking on the heart and circulatory system?

Take the pulse of another student who is a regular smoker. This should be done two or three times when the person is not smoking to establish a base line accuracy. Count the number of beats in a 15-second period and multiply by 4 to determine the number of pulsations per minute. Have the person light a cigarette, and after the third or fourth inhalation, take the pulse again.

You may wish to determine how many extra beats one pack of cigarettes causes a person based on the data you have gathered. Also, calculate any extra volume of blood pumped by the heart as a result of smoking one package of cigarettes. (Assume each heartbeat pumps about 70 ml of blood.)

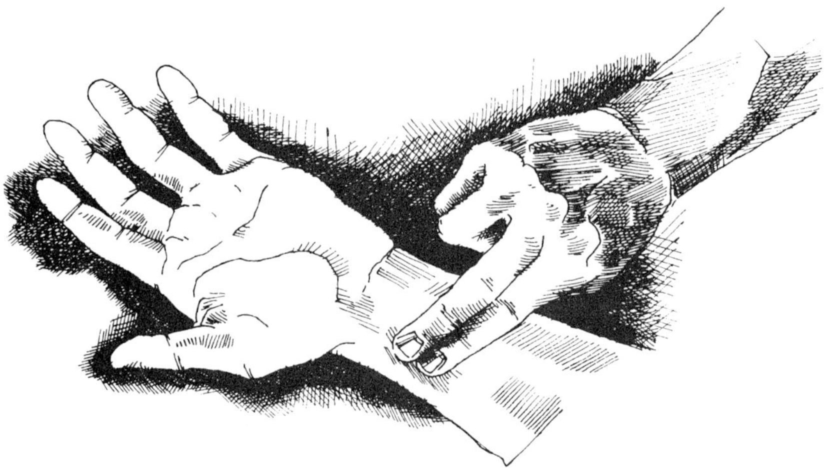

Figure 18. The arterial pulse taken at the wrist is an accurate indication of the heart rate. Take the pulse by placing the first and second fingers of your right hand on the thumb side of the person's wrist.

EXERCISE 5. What is the effect of smoking on body temperature?

Have a regular smoker who has not smoked for several hours hold a thermometer between thumb and forefinger. Record the highest temperature after a minute or two. Cool the thermometer by placing it in running tap water. Then have the smoker smoke and then hold the thermometer. Exactly as before, record the highest temperature.

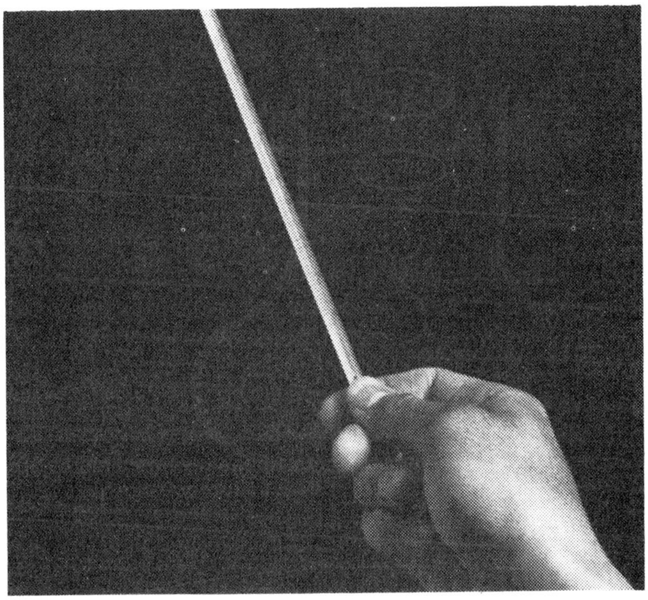

Figure 19. Holding the thermometer between thumb and forefinger will not give a true reading of body temperature but will give comparative readings of extremity temperatures, as before and after smoking.

2. What is the composition of cigarette smoke?

EXERCISE 6. Mechanical smoker

Directions for the operation of the mechanical smoker are at the central study area.

From your observations of the mechanical smoker, describe some of the substances of which cigarette smoke is composed.

cigarette smoke
- 92% gases
- 8% ash

tobacco smoke tar
- acidic—phenol
- neutral—benzopyrene
- basic—nicotine

Figure 20. Components of cigarette smoke.

3. Is smoking really dangerous to your health?

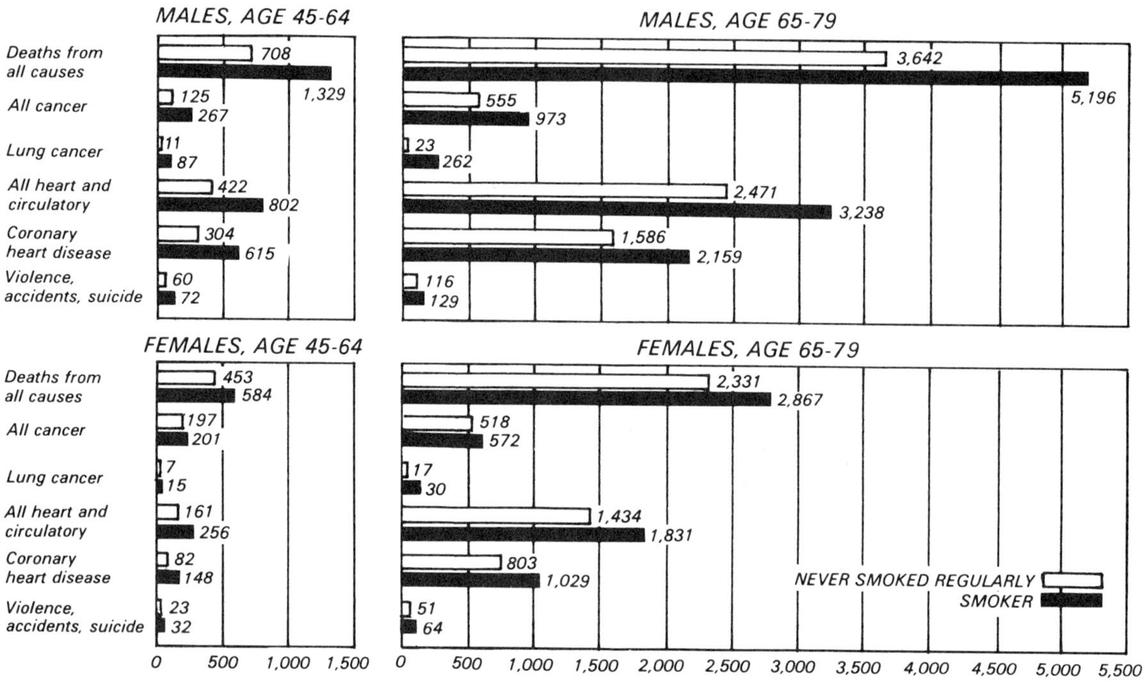

Figure 21. Death rates of smokers versus nonsmokers, 1972. (National Cancer Institute. Monograph No. 19 [Hammond Study]).

What kinds of evidence lead scientists to conclude that smoking is injurious to health?

a. animal experiments—exposure to smoke and tars

b. clinical and autopsy studies—observations on medical patients and autopsies of smokers and nonsmokers.

c. population studies

(1) retrospective—smoking histories of persons with a specific disease (lung cancer, for example) are compared with control groups of people having the disease but who do not smoke.

(2) prospective—smoking histories of smokers and nonsmokers are kept, and eventually cause of death is recorded when it occurs.

(3) clinical—the degree of occurrence of specific symptoms (coughing, lung function, and so on) is tabulated for both smokers and nonsmokers.

C. Marijuana

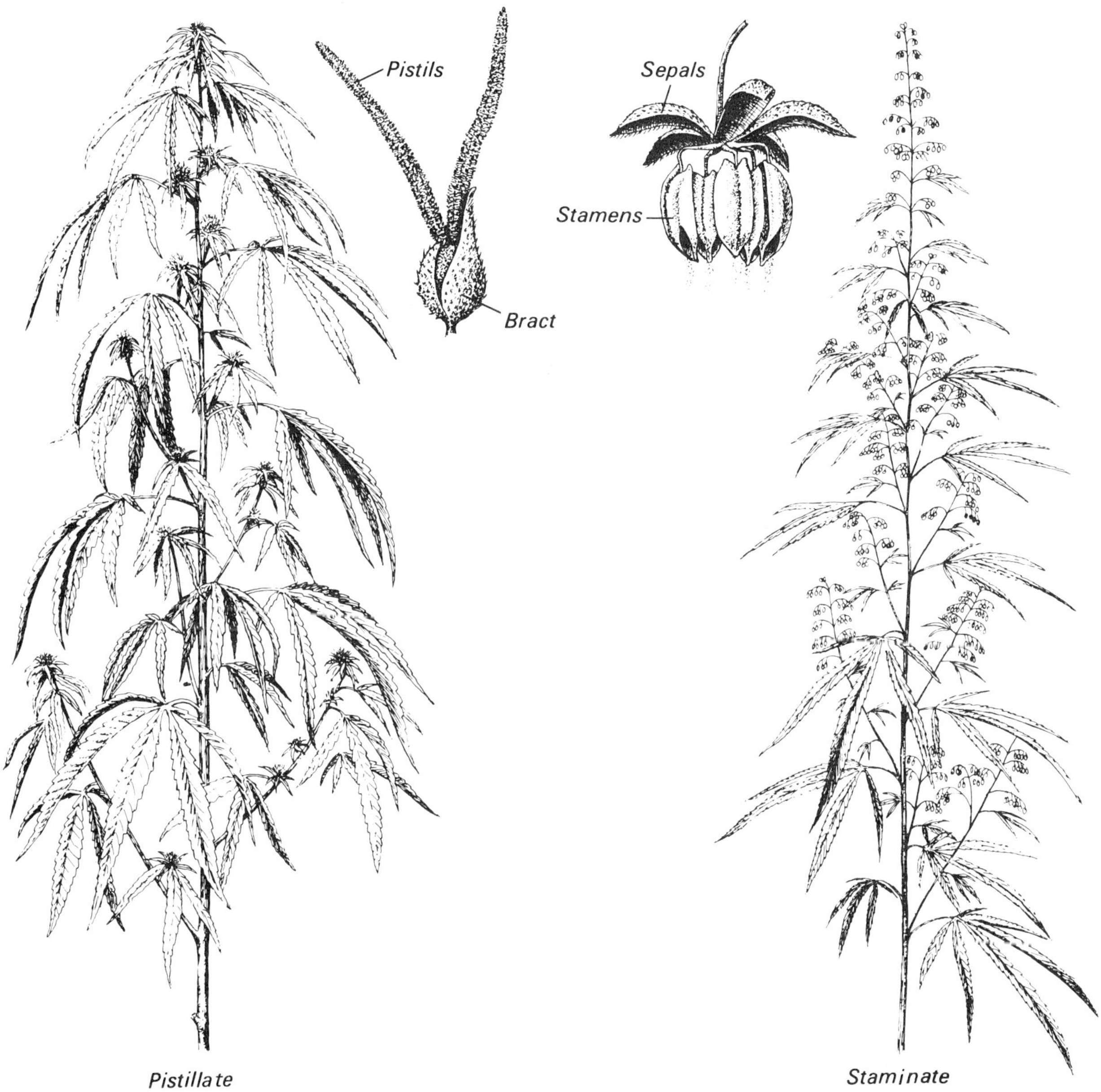

Figure 22. Marijuana plant (*Cannabis sativa*). The plant is dioecious, that is, the staminate flowers are on one plant (*right*) and the pistillate flowers on another (*left*). The most active drug substances are contained in a sticky yellow resin found on the flowers and top leaves of the female plant.

D. LSD—Lysergic Acid Diethylamide

From a report by Dr. Albert Hofmann, following an experience he had while working at a pharmaceutical company in Basel, Switzerland (1943):

On arriving home I lay down in a dazed condition with my eyes closed. There surged upon me an uninterrupted stream of fantastic images of extraordinary plasticity and vividness and accompanied by an intense, kaleidoscopelike play of colors. I felt a marked desire to laugh. I had great difficulty in speaking coherently, my field of vision swayed before me, and objects appeared distorted like images in curved mirrors. The faces of those around me appeared as grotesque, colored masks. I had a clear recognition of my condition, in which state I sometimes observed, in the manner of an independent, neutral observer, that I shouted half insanely or babbled incoherent words. Occasionally I felt as if I were out of my body. All acoustic perceptions (e.g., the noise of a passing car) were transformed into optical effects, every sound evoking a corresponding colored hallucination constantly changing in shape and color. . . .

VI. HOW IS THE PRODUCTION, DISTRIBUTION, AND USE OF DRUGS CONTROLLED IN OUR SOCIETY?

A. Problems of Drug Control

EXERCISE 7. Making a decision regarding drug control

The following true statements describe an actual drug:

a. This drug is widely used throughout the world, particularly in the U.S.

b. It is not known exactly how this drug works on the body and brain to produce its effects.

c. This drug has been found to have a causal relationship with birth defects in animals.

d. Thousands of children in the U.S. die or are disabled each year from accidental overdose of this drug.

Based on these facts, what restrictions do you believe should be placed on the manufacture, sale, and use of this drug?

B. Major U.S. Drug Legislation

1. 1906—Food and Drug Act
2. 1914—Harrison Narcotic Act
3. 1919–1933—Prohibition (16th Amendment)
4. 1937—Marijuana Tax Act
5. 1951—Boggs Act
6. 1956—Narcotic Drug Control Act
7. 1965—Drug Abuse Control Amendment
8. 1970—Comprehensive Drug Abuse Prevention and Control Act

C. Classes of Drugs (Including Examples) Whose Use, Possession for Sale, and Sale are Illegal

1. Class I—heroin, morphine, marijuana, LSD
2. Class II—cocaine, opium, medically used narcotics
3. Class III—codeine, amphetamines, barbiturates
4. Class IV—chloral hydrate, meprobamate, mild barbiturates
5. Class V—tranquilizers

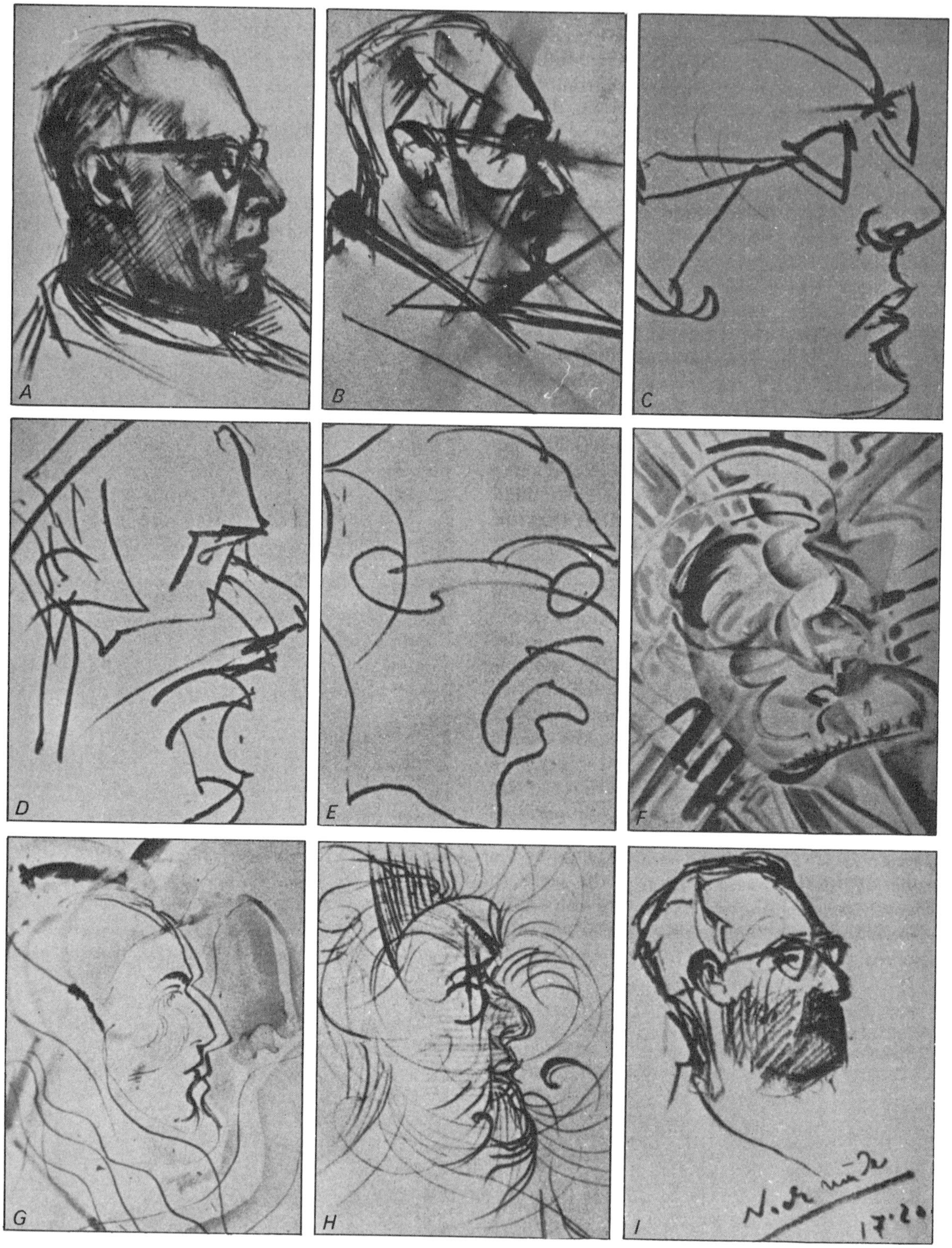

Figure 23. Drawings by an artist under the effects of LSD over a period of eight hours. (Reproduced from *Triangle,* the Sandoz Journal of Medical Science, Vol. 2, No. 3, 1955, pp. 119–123.)

APPENDIX I

Answers to Drug Information Survey

1. C. Habit-forming drugs differ from addictive drugs primarily in that the discontinuance of habit-forming drugs usually does not result in bodily hunger for its effects, or in significant physical distress.

2. D. While drug addiction is found in all socio-economic situations, statistical data confirm the greater frequency of addiction among people in city slums.

3. A. Barbiturates can effectively bring about sleep with little danger of addiction if used strictly according to medical prescription.

4. A. Such a baby may die if the addiction is not identified and treated.

5. A. Habitual use is related primarily to psychological and social drives reinforced by the action of nicotine on the central nervous system.

6. B.

7. C. In spite of the folklore among high school and college students, there is no scientific evidence that amphetamine use increases intellectual activity.

8. C. An amphetamine abuser is also prone to accidents that can be attributed to the drug and to the fact that fatigue has an effect in spite of the stimulation received from the drug.

9. D.

10. A. The effects of LSD are not quantitative. The effect on sight and vision is one perceived by the user.

11. In 1970 Americans spent about $15.7 billion for alcoholic beverages, $9 billion for tobacco, $3.2 billion for the caffeine beverages, coffee, tea, and cocoa, and $20 billion for aspirin. Although accurate figures are not available for the numbers of people who consume these substances, it is believed that aspirin is the most widely used.

12. alcohol

13. C. While marijuana does not appear at this time to be as harmful as some other drugs, such as alcohol, there are not enough data at the present time to justify sweeping conclusions concerning its effects.

14. B. No causal relationships have been established between marijuana use and use of other drugs. Sociological research supports B as the best answer.

15. B. While some persons can make occasional use of addictive and habit-forming drugs without becoming dependent upon them, there is no way to know in advance whether or not one will become dependent on a particular drug.

16. C. Anyone is susceptible to drug involvement. There is no way to predict proneness to drug abuse or likelihood of drug dependence.

17. C. A very real danger exists if there is abrupt withdrawal from a barbiturate for which strong physical dependence has been established.

18. A. Black market drugs are almost always cut or adulterated in a number of ways that may be harmful.

APPENDIX II

Dangerous Drugs Identification Kit for Educational Purposes

Stimulants—Amphetamines ("uppers")

1. Amphetamine Sulfate ("bennie," "bean," "white," or "cartwheel")
2. Methamphetamine Hydrochloride ("speed," "meth," or "crystals")
3. Dextroamphetamine Sulfate ("dexies," "hearts," or "peaches")

Depressants—Barbiturates ("downers")

4. Secobarbital Sodium ("reds," "redbirds," or "red devils")
5. Secobarbital Sodium in color variety ("pink lady," or "pinkie")
6. Sodium Pentobarbital ("yellow," or "yellow jacket")
7. Amobarbital Sodium and Secobarbital Sodium ("rainbow," or "double trouble")
8. Amobarbital Sodium ("blue," "blue heaven," or "blue velvet")

ADDITIONAL INQUIRIES

1. Investigate the current drug laws of your state. Identify what changes have occurred recently and what is presently under consideration for future change. Compare your state laws with those of some other states.
2. Visit a methadone clinic or drug rehabilitation center. Find out exactly what their functions are and whether there are positive results from the therapy provided.
3. Construct a carefully designed questionnaire that you can use to study the uses of and attitudes toward drugs.
4. Additional readings:

Brecher, E. M., and Consumer Report Editors. 1973. Licit and illicit drugs: the consumer union report on narcotics, stimulants, depressants, inhalants, hallucinogens, and marijuana—including caffeine, nicotine, and alcohol. Little, Brown and Company, New York.

Fort, J., M.D. 1972. The drug explosion. Playboy Magazine. *19*(9).

Ray, Oakley S. 1972. Drugs, society, and human behavior. C. V. Mosby Company, St. Louis.

Shepherd, Jack. 1974. Smack city. Harpers. *248*(1485):32–34.

Weinswig, Melvin. 1973. Use and misuse of drugs subject to abuse. Pegasus, a division of Bobbs-Merrill, Inc., New York.

GLOSSARY

This glossary includes some terms that were not used in this minicourse but which are commonly used in drug terminology.

ANALGESIC: A drug that produces relief from pain.

ANESTHETIC: A drug that causes a loss of sensation or feeling, or loss of consciousness.

ANTIDEPRESSANT: A drug used to relieve or prevent depression.

BASAL GANGLIA: An area of clustered interconnected nerve cells located in the inner part of the cerebrum.

DEPRESSANT: A drug that reduces bodily functional activity.

DRUG: Any biologically active substance that is deliberately introduced into the body to affect its functioning.

HALLUCINOGEN: A drug that induces sensation not based on reality (psychedelic, psychotomimetic).

HYPNOTIC: A drug that induces sleep.

MOTOR NEURON: A nerve that carries impulses away from the brain and spinal cord to muscles and glands.

MULTIHABITUATION: The use of two or more drugs interchangeably in an attempt to counteract the adverse effects of one or the other, or to potentiate the effects of one or the other.

NARCOTIC ANALGESIC: Opium, its derivatives, and other compounds with actions similar to opium. These compounds relieve major pain.

NEUROTRANSMITTER: A chemical substance released from the end of a nerve fiber (axon) into a synapse, from which it affects nearby structures (e.g., other nerves, muscles, glands).

PAN-ADDICTION: Physical dependence on two drugs of different classes.

PHARMACOLOGICAL: Pertaining to the properties and reactions of drugs.

PHYSIOLOGICAL DEPENDENCE: Addiction—a state characterized by tolerance and withdrawal.

PSYCHOLOGICAL DEPENDENCE: Habituation—a need or craving for a drug but without withdrawal effects.

RETICULAR FORMATION: A network of ascending and descending fibers in the midbrain, which acts as a sensory filter for incoming information.

SEDATIVE: A drug that reduces tension and anxiety (calms); its use is often accompanied by excessive drowsiness.

SENSORY NEURON: A nerve that carries impulses from the point of stimulus (reception) to the spinal cord and brain.

STIMULANT: A drug that induces excitement and stimulation.

SYNAPSE: The junction between a nerve cell and another cell.

TOLERANCE: Gradually diminishing effect of a drug as a result of repeated administration of a given dose of the drug.

TRANQUILIZER: A drug whose primary effect is to reduce tension and anxiety (calms), generally without inducing excessive drowsiness.

WITHDRAWAL: A series of characteristic symptoms that occur upon abrupt cessation of the use of the drug.

REFERENCES

Adams, E. 1958. Barbiturates. Scientific American. *198*(1):60–64.

Alsop, S. 1971. I didn't stop in Baltimore. Newsweek. *83*:92.

Barber, B. 1967. Drugs and society. Russel Sage Foundation, New York.

Barron, F., et al. 1964. The hallucinogenic drugs. Scientific American. *210*(4):29–37.

Blum, R. H., et al. 1969. Society and drugs. Jossey Bass, San Francisco.

Brecher, E. M., and Consumer Reports Editors. 1973. Licit and illicit drugs: the consumer union report on narcotics, stimulants, depressants, inhalants, hallucinogens, and marijuana—including caffeine, nicotine, and alcohol. Little Brown and Company, Boston.

Byrd, O. E. 1970. Medical readings on drug abuse. Addison-Wesley Publishing Company, Reading, Massachusetts. p. 69.

Campbell, R. 1971. The chemistry of madness. Life. *71*:66–86.

Chemical and Engineering News. 1970. Pharmacology of marijuana remains obscure. Chemical and Engineering News. October 26, 1970, p. 36.

Clark, M. E. 1973. Contemporary biology: concepts and implications. W. B. Saunders Company, Philadelphia.

Cohen, S. 1969. The drug dilemma. McGraw-Hill Book Company, New York.

Communication Research Machines, Inc. 1972. Life and health. Ziff-Davis Publishing Company, Del Mar, California. pp. 117–210.

Fort, J. 1969. The pleasure seekers. The Bobbs-Merrill Company, Inc., Indianapolis, Indiana.

Grinspoon, L. 1969. Marijuana. Scientific American. *221*(6):17–25.

Hu, C.-Y. 1974. Effects of ethylene in cigarette smoke on the development of pea seedlings. American Biology Teacher. *36*(4):233–235.

Johnson, C. E. 1967. Mystical force of the nightshade. International Journal of Neuropsychiatry. *3*:274.

Kaplan, J. 1971. Marijuana—the new prohibition. World Publishing Company, New York.

Kogan, B. A. 1970. Health: man in a changing environment. Harcourt Brace Jovanovich, Inc., New York.

Lieberman, C. M., and B. W. Lieberman. 1971. Marijuana—a medical review. New England Journal of Medicine. *284*:88–91.

National Clearinghouse for Drug Abuse Information. 1971. Drug dependence and abuse: a selected bibliography. Superintendent of Documents, U.S. Government Printing Office, Washington, D.C.

Ray, O. S. 1972. Drugs, society, and human behavior. C. V. Mosby Company, St. Louis.

Rech, R. H., and K. E. Moore. 1971. An introduction to psychopharmacology. Raven Press, New York.

Science Service, Inc. 1972. The drugged Americans: aspirin poppers not spared. Science News. *101*:7–8.

Snyder, S. H. 1971. Work with marijuana. I. Effects. Psychology Today, Vol. 5.

Tart, Charles. 1971. Work with marijuana. II. Sensations. Psychology Today, Vol. 5.

Time, Inc. 1973. The deadly downer. Time. March 5, 1973, pp. 73–74.

Time, Inc. 1974. Livers and liquor. Time. *103*:61.

U.S. Department of Health Education and Welfare. 1973. Chartbook on smoking, tobacco and health. #(HSM)73-8718. Superintendent of Documents, U.S. Government Printing Office, Washington, D.C.

U.S. Department of Justice, Drug Enforcement Administration. Drugs of abuse. #2704-0017. Superintendent of Documents, U.S. Government Printing Office, Washington, D.C.

Weil, A., et al. 1968. Clinical and psychological effects of marijuana in man. Science. *162*:1234–1242.

Weinswig, M. H. 1973. Use and misuse of drugs subject to abuse. Pegasus, a division of Bobbs-Merrill Company, Inc., New York.